RFID Explained: A Primer on Radio Frequency Identification Technologies

A Lecture on Radio Frequency Identification
Roy Want

ISBN: 978-3-031-01346-1   paperback
ISBN: 978-3-031-02474-0   ebook

DOI: 10.1007/978-3-031-02474-0

A Publication in the Springer series
SYNTHESIS LECTURES IN MOBILE AND PERVASIVE COMPUTING #1

Lecture #1
Series Editor: Mahadev Satyanarayanan, Carnegie Mellon University

First Edition
10 9 8 7 6 5 4 3 2 1

RFID Explained: A Primer on Radio Frequency Identification Technologies

Roy Want
Intel Research

ABSTRACT

This lecture provides an introduction to Radio Frequency Identification (RFID), a technology enabling automatic identification of objects at a distance without requiring line-of-sight. Electronic tagging can be divided into technologies that have a power source (active tags), and those that are powered by the tag interrogation signal (passive tags); the focus here is on passive tags. An overview of the principles of the technology divides passive tags into devices that use either near field or far field coupling to communicate with a tag reader. The strengths and weaknesses of the approaches are considered, along with the standards that have been put in place by ISO and EPCGlobal to promote interoperability and the ubiquitous adoption of the technology. A section of the lecture has been dedicated to the principles of reading co-located tags, as this represents a significant challenge for a technology that may one day be able to automatically identify all of the items in your shopping cart in a just few seconds. In fact, RFID applications are already quite extensive and this lecture classifies the primary uses. Some variants of modern RFID can also be integrated with sensors enabling the technology to be extended to measure parameters in the local environment, such as temperature & pressure. The uses and applications of RFID sensors are further described and classified. Later we examine important lessons surrounding the deployment of RFID for the Wal-Mart and the Metro AG store experiences, along with deployments in some more exploratory settings. Extensions of RFID that make use of read/write memory integrated with the tag are also discussed, in particular looking at novel near term opportunities. Privacy and social implications surrounding the use of RFID inspire recurring debates whenever there is discussion of large scale deployment; we examine the pros and cons of the issues and approaches for mitigating the problems. Finally, the remaining challenges of RFID are considered and we look to the future possibilities for the technology.

KEYWORDS

Automatic identification, distributed memory, electronic tagging, passive tagging, privacy debate, radio frequency Identification (RFID), remote sensing

Contents

List of Figures

Acknowledgements

The author would like to thank Satya for providing the opportunity to write about RFID for the Morgan Claypool Synthesis Lecture Series, and Waylon Brunette, Adam Rea, Gaetano Borriello, Trevor Pering, Josh Smith, and his family Susan Want and daughters Hannah and Becky for their constant support.

CHAPTER 1

Introduction

Ever since the advent of large-scale manufacturing, rapid identification techniques have been needed to speed the handling of goods and materials. Historically, printed labels, which are a simple cost-effective technology, have been the staple of the manufacturing industry. In the 1970s, labeling made a giant leap forward with the introduction of UPC barcodes [1] making it possible to both automate and standardize the identification process. Barcodes [2], although very inexpensive to produce, have many limitations: a clear line of sight is needed between the reader and the tag, they can be obscured by grease and nearby objects, are hard to read in sunlight or when printed on some substrates (Fig. 1.1). There are many types of barcode in use for specialty applications, including block-based optical codes. There are even miniture plastic barcodes called taggants [3] incorporated into explosives that are design to withstand an explosion and identify the supplier in case used for criminal purposes. However, for the scope of this article, barcodes labels and optical codes are considered as a single group. An alternative labeling technology is Radio Frequency Identification (RFID) [4–6], which enables identification at a distance without a line of sight. To provide some context Fig. 1.2 shows a variety of RFID tags designed for diverse applications in comparison to the size of a dime. Figure 1.3 shows a typical tag reader and remote antenna that can be installed in the area the tags are expected to pass through. It should be noted that as RFID is a radio technology, the tags do not need to be visible at all, and can be concealed behind an aesthetically designed label, or even molded into the product housing itself. For these reasons it is possible that many people may have encountered RFID tags but were unaware of them due to their invisible placement. Electronic tagging is superior to barcodes in many ways as it can reliably support a much larger set of unique IDs, and incorporate additional data, such as the manufacturer, and product serial number (Fig. 1.4). Furthermore, RFID systems can discern many different tags that are located in the same general area without human assistance. In contrast, consider the individual care needed at a supermarket checkout where each item is carefully orientated toward the reader before it can be scanned.

RFID is not a new technology. For example, the principles of RFID were employed by the British in World War II to identify their aircraft using the IFF system (Identity: Friend or Foe). Later, work on access-control that is more closely related to modern RFID, was carried out at Los Alamos National laboratories during the 1960s. In this application, RFID tags incorporated

FIGURE 1.1: Examples of (a) 1D (39 Code) and (b) 2D barcodes

FIGURE 1.2: RFID tags—various shapes and sizes

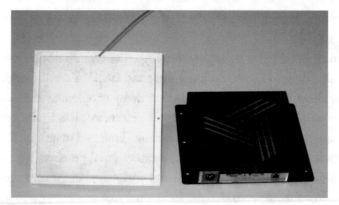

FIGURE 1.3: RFID reader and tag

Property	Barcodes	RFID
Line of sight required	Yes	No
Can be read in sunlight	No	Yes
Needs to be oriented to be read (operator involved)	Yes	No
Affected by grease and grim	Yes	No
Cost in volume (greater 1 M units)	Free (printed)	10–12 cents (in 2006)
Aesthetic integration with product	No	Yes
Typical Number of ID bits	1D: 80 bits 2D: max 2 kbits	96 bits (EPC) w/Memory up to 8 kbits
Processing function options	No	Yes (crypto, hash)
Additional memory	No	8 kbits (current max)
In-situ read/write capability	Read only	Yes
Multi-tag arbitration	No	Yes
Disable option at POS	No	Yes

FIGURE 1.4: Comparing barcodes and RFID

in employee badges enabled automatic identification of people to limit access to secure areas, and had the additional advantage that it made the badges hard to forge. For many years this technology has been relatively obscure, although it has been adopted in various niche domains, such as to identify animals, make toys interactive, improve car-key designs, label airline luggage, time marathon runners, prevent theft, enable automatic toll-way billing, and many forms of ID badge for access control. Today, it is even being applied to validate money and passports, and as a tamper safeguard for product packaging.

In recent years, RFID has been widely written about, and even appeared in a primetime television advertisement as a promotion of IBM's business solutions. The technology has moved from obscurity to applications that are now firmly in the public eye, and the ethics of its use are regularly debated by journalists, technologists, and privacy advocates. You may ask why it took over 50 years for the technology to become mainstream? The primary reason is one of cost. When electronic identification technologies compete with the rock bottom pricing of printed symbols on paper, it either needs to be equally low-cost, or provide enough added value to an organization that the cost is recovered elsewhere. RFID is now at a critical price-point that could enable its large-scale adoption for the management of consumer retail goods. At the time of writing this article, Alien Technologies [7] are able to supply a modern RFID tag at a unit price of 12.9 cents in quantities of 1 million, notably still quite a bit more expensive than printed sticky labels. When adoption begins to take hold, it will rapidly accelerate as volume production drives prices down, making it more attractive to deploy the technology to support a wider range of markets. Modern semiconductor manufacturing has also played a role in the progress of RFID design, driving up the functionality of the tags, at progressively lower power, using a smaller area of silicon, which in turn lowers cost. The sensitivity of the receiver in the reader has also increased and these improvements in the reader and tag can now be achieved at low manufacturing costs (see Section 2).

Today, applications of RFID are being extended to new domains. The European Union is considering putting RFID into ECUs; a small chip that is sandwiched between the layered papers of the European paper denomination. The objective is to make forgeries more difficult and provide automatic tracing of its use—although in the latter case, some people feel this undermines the benefits afforded by paper money.

On a similar theme, the US government is also planning to incorporate RFID into the US passports to reduce counterfeiting and enable efficient automatic checks at the national border.

RFID is also being used to tag family pets, and in some states a dog registration process involves injecting a sub-dermal tag under the skin of the dog. Unlike a neck collar the tag cannot be accidentally removed, or fall off, and even intentional theft of the animal requires surgical skills if its identity is to be obscured. RFID is already responsible for reuniting many runaway dogs with their owners.

There are three primary organizations that are pioneering the adoption of RFID on a large scale: Walmart, Tesco, and the US DoD. Each is driven by the potential to lower their operational costs in order to have the most competitive product pricing. RFID, with its fast-read times and high-reliability can do this by streamlining the tracking of stock, sales, and orders. When used in combination with computerized databases and inventory control, linked together by digital communication networks across the globe, it is possible to pinpoint the progress of individual items between factories, warehouses, transportation vehicles, and stores.

The potential benefits of automatic RFID tracking yielding improvements in efficiency are alluring to large companies that are trying to squeeze the cost out of manufacturing, distribution, and retail within their organization. The economics of this attraction will be a major force in the adoption of the technology and will also drive improvements in its own evolution through the resulting investment. It is interesting to note that the announcements of plans to roll-out electronic tagging have also stirred up concerns that personal privacy may be eroded. RFID tagging opens up the possibility for item level identification, and that means products that we buy and carry with us, which contain RFID, can uniquely identify us, and further more this can be done covertly at a distance. The resulting privacy implications are discussed in a later section.

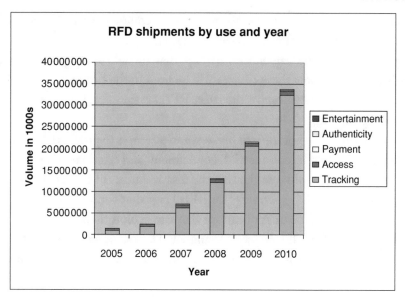

FIGURE 1.5: Forecast volume RFID tags shipped 2005–2010 (1000s). Source: Instat 12/05

To provide an overall picture of how RFID is expected to be adopted in the coming years see Fig. 1.5, and In-stats forecast for RFID adoption [8].

The expected growth curve is the classic hockey-stick shape with tag sales estimated at 1.3 billion in 2005, and expected to rise to 33 billion by 2010, a 2500% increase over in 5 years. This is a phenomenal growth by any standard. The 2010 forecast expects tag applications to be dominated by tracking at 94%, which can be broken down as supply chain (83%) and food (11%). Access control (3%) is the next highest volume use and all other uses are expected to total 3%. From a financial perspective, revenues are expected to increase from $683M to $2480M in the same time frame.

1.1 LECTURE OVERVIEW

In this section we have provided an overview of the broad topic of RFID. The following sections examine the technology and its implications in greater depth, and are organized as follows:

Section 2 describes the *principles* of RFID operation, including the various types of tagging system. We consider the physics behind the design of RFID and the physical constraints that limit how tags can be used.

Section 3 looks at the most influential *standards* that are shaping the adoption of electronic tagging technologies across the industry.

Section 4 examines the issues associated with reading multiple *collocated tags* and the algorithms that can be used to resolve them.

Section 5 provides a more in-depth overview of commercial *applications* by classifying them and providing detailed examples in each category.

Section 6 describes the capabilities of tags that include *sensing*, some of the existing products in this space and how these devices will be used in the future.

Section 7 presents *deployments* of the latest RFID systems and summarizes the key *experiences* found along the way.

Section 8 considers the hot topic of *privacy*, and attempts to provide a broad consideration of how RFID might be used by commercial, government, and criminal groups, and how we can benefit from the advantages while mitigating the disadvantages of the technology.

Section 9 introduces the *opportunity* for RFID tags that contain read/write *memory*, their memory capacity and the scope of applications that can exploit them.

Section 10 summarizes the *challenges* facing RFID, from several perspectives: technology, manufacturing cost, deployment, and social acceptance; and concludes by looking at the *future* of electronic tagging.

CHAPTER 2

Principles of Radio Frequency Identification

There are many types of RFID [9], and at the highest level of classification these can be divided into two classes: *active* and *passive* devices. Active tags require a power source [10] and either need a connection to powered infrastructure or have a limited lifetime defined by the energy stored in an integrated battery, balanced against the number of read operations that will be performed on the tag. Examples of active tags are, transponders attached to aircraft to identify their national origin, and LoJack devices attached to cars that incorporate cellular technology along with a Global Positioning System (GPS), communicating the location of a car if stolen. Olivetti Research Ltd's Active Badge, used to determine the location of people and objects in a building is an example of a small wearable active tag with a lifetime of about 1 year [11]. There are also some types of active tag that scavenge power from their enviornment. MIT Media lab's push-button powered doorbell controller [12] is another; the mechanical energy scavenged from pushing the switch is used to power the electronics.

However, it is the passive RFID tag that is of interest to retailers, requiring no maintenance and exhibiting an indefinite operational life [13–16]. They have no battery, and can be made small enough to be incorporated into a practical adhesive label. A passive tag consists of three parts: an antenna, a semiconductor chip attached to the antenna, and some form of encapsulation, which could be a small glass vial or a laminar plastic substrate with adhesive on one side to enable easy attachment to goods, see Fig. 2.1. The encapsulation is necessary to maintain the integrity of the tag and protect the antenna and the chip from environmental conditions or reagents that would cause damage.

The purpose of the tag antenna is to receive power from the reader, and shortly after to transmit its ID in response. The tag chip is powered by the energy in the signal received at the tag antenna, which activates an electronic circuit and encodes an ID onto the return signal that, in turn, is communicated back to the reader by the antenna. In the history of RFID design there have been two fundamentally different design approaches for delivering power from reader to tag: magnetic induction and electromagnetic wave capture. These two designs take advantage of the electromagnetic properties associated with an RF antenna; the *near field* and the *far field*.

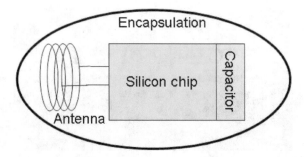

FIGURE 2.1: Logical components of an RFID tag. Note that the antenna can take many forms including a coil and a dipole depending on the tag type

If an alternating current is passed through a coil it will create an alternating magnetic field in the locality of the coil, and this is referred to as the *near field*. This circuit will also give rise to propagating electromagnetic waves that breakaway from the coil/antenna and radiate into space, this is termed the *far field* and is the principle of radio transmission. Both radio properties can be used to transfer enough power to a remote tag to sustain its operation, typically between 10 µW and 1 mW depending on the tag type, and through modulation can also transmit and receive data [4]. To show how small these power budgets are, by comparison, the nominal power consumed by an Intel XScale processor is approximately 500 mW and an Intel Pentium-4 is 50 W.

2.1   NEAR-FIELD-BASED RFID DESIGN

The use of near-field coupling between reader and tag can be described in terms of Faraday's principle of magnetic induction. A reader passes a large alternating current through a reading coil, resulting in an alternating magnetic field in its locality. If a tag that incorporates a smaller coil (Fig. 2.2) is placed in this field, an alternating voltage will appear across it, and if rectified and coupled to a capacitor, a reservoir of charge will accumulate that can be used to power a tag chip. Tags that use near-field coupling send data back to the reader using *load modulation*. Since any current drawn from the tag coil will give rise to its own small magnetic field which will oppose the reader's field, this can be detected at the reader coil as a small increase in current flowing through it. This current is proportional to the load applied to the tag's coil (hence *load modulation*), and is the same principle used in power transformers found in most homes today—although usually the primary and secondary coil of a transformer are wound closely together to ensure efficient power transfer. Thus, if the tag's electronics applies a load to its own antenna coil and varies it over time, a signal can be created that encodes the tag's ID, and the

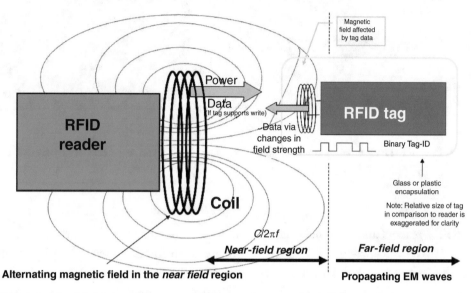

Using induction for power coupling from reader to tag
and load modulation to transfer data from tag to reader

FIGURE 2.2: Near-field power/communication mechanism for RFID tags operating at less than 100 MHz

reader can recover this signal by monitoring the change in current through the reader coil. A variety of modulation encodings are possible depending on the number of bits of ID required, the rate of data transfer, and additional redundancy bits placed in the code to remove errors resulting from noise in the communication channel.

Near-field coupling is the most straight forward approach for implementing a passive RFID system and as a result it was the first, and has led to many standards such as ISO 15693 and 14443 (see Section 3), and a variety of proprietary solutions. However, near-field communication has some physical limitations. It turns out that the range for which it is possible to use magnetic induction approximates to $c/2\pi f$. Thus, as the frequency (f) of operation increases, the distance that near-field coupling can operate over decreases (c being a constant, the speed of light). A further limitation is the energy available for induction as a function of distance from the reader coil. It can be shown (Section 2.2 below) that the magnetic field drops off at a $1/x^3$ factor along a center line perpendicular to the plane of the coil. As applications require more ID bits, and have the requirement to discriminate between multiple tags in the same locality during a fixed read time, it is necessary to increase the data rate used by the tag and thus the operating frequency. These design pressures have led to new passive RFID designs based on far-field communication (Fig. 2.3).

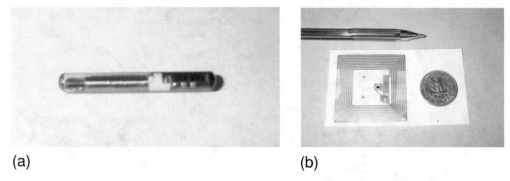

(a)   (b)

FIGURE 2.3: RFID tags based on near-field coupling. (a) Trovan tag (128 kHz), size: 1 cm [17]. (b) Tiris (13.56 MHz), size: 5 cm × 5 cm [18]

2.2   PROPERTIES OF MAGNETIC FIELDS

Since RFID readers and near-field tags couple to each other using Faraday's principle of induction [19], we can analyze the basic interaction between RFID reader and tag by considering the two devices as two circular coils parallel to each other and aligned along a perpendicular line that runs through their centers. This is a similar configuration to the more familiar AC power transformer. If current is passed through the reader coil, the magnetic field H (in Webers) will be defined at its center by (I current in Amperes, N number of turns, r radiues of coil):

$$H = IN/2r$$

And the magnetic flux density B (for free space) is given by:

$$B = \mu_0 H$$

The resulting characteristic magnetic field pattern around the coil is shown below in Fig. 2.4. As you can see in the center of the coil the magnetic field is perpendicular to the plane of the coil and extends outward along that axis becoming weaker at a distance. In comparison,

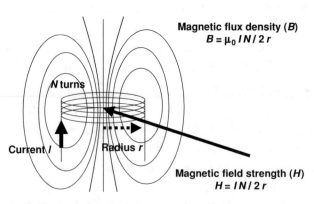

FIGURE 2.4: Magnetic field calculation at the center of a coil

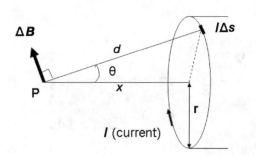

Using Biot-Savart law for calculating the magnetic flux density at P by considering the sum of current elements $I\Delta s$

$$B = \Sigma \, (\mu_0/4\pi) \cdot I\Delta s \cdot \sin\theta \,/\, d^2$$

$$B = (\mu_0/4\pi) \cdot 2 \, \pi \, r \, N \, I \cdot \sin\theta \,/\, d^2$$

$$B = (\mu_0 \cdot I \cdot N \,/\, 2) \cdot r^2 \,/\, (r^2 + x^2)^{3/2}$$

FIGURE 2.5: Magnetic flux density (B) at a distance x from the center of an N-turn coil, with radius r, and current flowing, I

a power transformer has its primary and secondary coil tightly coupled and therefore the field strength at a distance is not normally of interest.

However, in the case of an RFID tag, it is the projection of the reader's magnetic field that allows a reader to interact with a tag at a distance even with optically opaque materials in the path. To understand how the field strength varies along this line we can consider the equations in Fig. 2.5.

Since the distance x from the center of the coil will be large compared to the radius of the coil, it will dominate the magnitude of B, which can be approximated to $1/x^3$, which is a rapidly decreasing function and one of the main reasons why near-field coupling, using practical field strengths, can typically only be used to read tags at up to 1 m away, see Fig. 2.5.

2.3   FAR-FIELD-BASED RFID DESIGN

RFID tags based on far-field coupling (Fig. 2.6) capture electromagnetic waves propagating from a dipole antenna attached to the reader. A smaller dipole antenna in the tag will receive

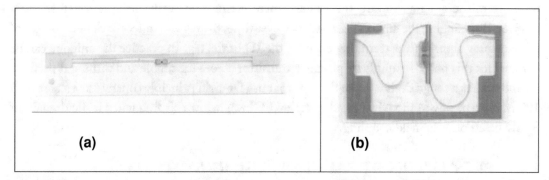

(a)   **(b)**

FIGURE 2.6: RFID tags based on far-field coupling. (a) Alien (900 MHz), size: 16 cm × 1 cm. (b) Alien (2.45 GHz), size: 8 cm × 5 cm

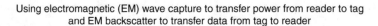

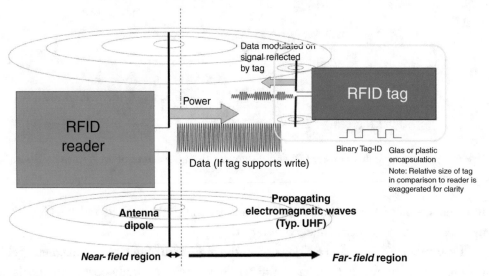

FIGURE 2.7: Far-field power/communication mechanism for RFID tags operating at greater than 100 MHz

this energy as an alternating potential difference that appears across the components of the dipole. This signal can also be rectified and used to accumulate energy in a capacitor reservoir to power its electronics. However, unlike the inductive designs, these tags will be beyond the range of the reader's near field, and information cannot be transmitted back to the reader using load modulation. The technique used by commercial far-field RFID tag designs is *back-scattering* (Fig. 2.7). If an antenna is designed with precise dimensions, it can be tuned to a particular frequency band and absorb most of the energy that reaches it in that band. However, if there is an impedance mismatch at this frequency, some of this energy will be reflected back as tiny waves from the antenna toward the reader, where it can be detected using a sensitive radio receiver. By changing the antenna's impedance over time, the tag can reflect back more or less of the incoming signal in a pattern that encodes the ID of the tag. In practice the antenna can be detuned for this purpose simply by placing a transistor across the dipole and turning it partially on and off. As a rough design guide the tags that use far-field principles operate at greater than 100 MHz typically in the UHF band (e.g., 2.45 GHz); below this frequency is the domain of RFID based on near-field coupling.

2.4   PROPERTIES OF BACKSCATTER RF SYSTEMS

The range of a far-field system is limited by the amount of energy that reaches the tag from the reader, and the sensitivity of the reader's radio receiver to the reflected signal. The actual

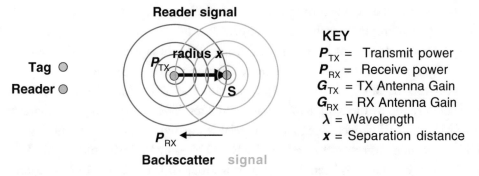

Reader signal

Tag ○
Reader ○

KEY
P_{TX} = Transmit power
P_{RX} = Receive power
G_{TX} = TX Antenna Gain
G_{RX} = RX Antenna Gain
λ = Wavelength
x = Separation distance

Backscatter signal

Transmitted power density at distance r in far-field
$$S = P_{TX}.G_{TX}/4\pi.X^2$$
Reflected power back to reader from a tag
$$P_{RX} = P_{TX}.G_{TX}.G_{RX}\,\lambda/(4\pi)^4.X^4$$

Doubling separation of tag and reader requires x16 more
transmit power to maintain the power level in the response

FIGURE 2.8: Backscatter signal strength at a distance x

return signal is very small because it is the result of two inverse square laws, the first as EM waves radiate from the reader to the tag, the second when reflected waves travel back from the tag to the reader (see Fig. 2.8). Thus, the returning energy is proportional to $1/x^4$ where x is the separation of the tag and reader.

As can be seen from the power density equations above, the absolute receive power is also proportional to the product of the transmitter and receiver's gain (G), and the wavelength of the carrier signal. Fortunately, thanks to Moore's Law [20], and the shrinking feature size of semiconductor manufacturing, the energy required to power a tag at a given frequency of operation continues to decrease (as low as a few microwatts). Moore's Law is more usually used to explain the increasing speed of computers for each new generation of semiconductor. However, if a CMOS transistor is built with reduced dimensions, at a given switching frequency it will consume less power. Thus, RFID tags build from CMOS transistors can operate at lower power when manufactured using smaller lithographies, and this has the consequence of increasing the operational range. In step with this trend, the sensitivity of inexpensive radio receivers has also been improving, and can now detect signals with power levels on the order of −100 dBm in the 2.4 GHz band. A typical far-field reader can successfully interrogate tags 3 m away, and some RFID companies claim their products have read ranges of up to 6 m. This is the result of many factors that include improved component tolerances, better antenna design, low-noise transistors, improved tag signal coding along with signal processing at the receiver to decode data on the return signal. Modern coding techniques support this trend by allowing more bits to be coded per cycle of the carrier. Moore's law plays an indirect role in this broad evolution by enabling inexpensive high-performance processors in the reader to run complex signal-processing algorithms in real time. Putting all of these factors together, RFID tag and reader designs can now be built that are more effective than ever before.

The work of EPCglobal Inc. [13], (originally the "MIT Auto-ID Center," a nonprofit organization set up by the MIT Media Lab., and later divided into Auto-ID laboratories, still part of MIT, and EPCglobal Inc., a commercial company), was key to promoting the design of UHF tags which has been the basis of the RFID trials by Walmart and Tesco. Although an extensible range of tags has been defined by this group, it is the Class-1 Generation-1 96-bit tag that has been the focus of recent attention. It has the flexibility for an EPC manufacturer to create over 12×10^{17} codes, making it possible to uniquely label every manufactured item for the foreseeable future, and not just using generic product codes. While this is not necessary for basic inventory control, it does have implications for tracing manufacturing faults, stolen goods, detecting forgery, and for the more controversial postsale marketing opportunities, enabling directed-marketing based on prior purchases (Section 8).

2.5   MODULATION TECHNIQUES USED WITH RFID

RFID tags that are designed to use backscatter have limited options for modulating data sent back to the reader. Amplitude Shift Keying (ASK) is the most basic and easiest to implement, but like all amplitude modulation techniques, this approach is prone to the affects of channel noise. When load modulation is used to transmit information, there are a greater variety of modulation options. Phase Shift Keying (PSK) is more robust than ASK, and in some designs

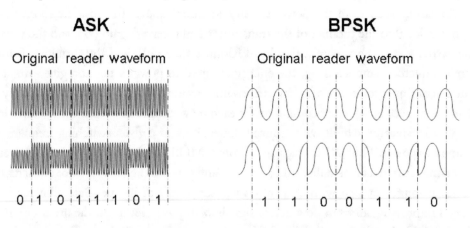

FIGURE 2.9: Modulation coding options for RFID signaling

Property	LF 125–135 kHz	HF 13.56 MHz	UHF 868,902–915 MHz	UHF 2.45 GHz
Typical max read distance (relates to antenna size)	1–2 m	2 m	20 m	20 m
Blocked by people (water)	No	No	No	Yes
Products available (typ.)	40+ years	10 years	Gen-1 4 years Gen-2 <1year	4 years
Tag powered by	Induction	Induction	EM capture	EM capture
Tag to reader comms.	Load modulation	Load modulation	Backscatter	Backscatter
Example vendor	Trovan, Tiris	Tiris (TI)	Impinj, Alien	Alien
Relative antenna size	Small	Medium	Large	Large
Data read rate	100 bps	2 kbps	Gen-1: 140–kbps Gen-2: 40–640 kbps	40 kbps
Multitag reads (tags/s)	N/A	30	Gen-1 500 Gen-2 1500	500

FIGURE 2.10: Comparing the properties of RFID operating in different frequency bands

a Binary PSK approach has been used successfully. Figure 2.9 shows how a tag transforms the received reader signal in order to send data back to the reader.

2.6   COMPARISON OF THE PROPERTIES OF RFID BASED ON FREQUENCY

Given the highly diverse properties of RFID tags described in this section it is useful to summarize by comparing the typical characteristics of the tags based on frequency. Figure 2.10 groups tags into Low Frequency (LF), High Frequency (HF), and Ultra High Frequency (UHF)—both 900 MHz and 2.45 GHz—and catalogs their typical characteristics. However, it should be emphasized that within each category the properties can vary considerably by manufacturer and application. For example, the antenna size used in the reader and the tag design, respectively, will have considerable bearing on the read range. For this reason it is unwise to project some trends from the table, for instance it appears that larger antennas are required for the high-frequency tags. However, this is not the case as the larger antenna is needed to achieve the long read-range that is possible with backscatter modulation. A UHF antenna could be made very small if it were only expected to be read from a few centimeters away. But as a general guide to the practical implementations of RFID technologies, the table serves its purpose.

CHAPTER 3

RFID Industry Standards

The International Standards Organization (ISO) and European Telecommunication Standards Institute (ETSI) have been set up to establish industry-wide standards across many disciplines. RFID can directly benefit from standardization to ensure widespread interoperability, and industry wide adoption has already been enabled by standards in common use; examples include ISO15693 and ISO14443.

There have been many RFID standards created over the years, each designed to solve a particular set of application requirements. The technology available to implement various standards has also improved over time and hence older standards define capabilities that are no longer state-of-the-art. Furthermore, the standards are very different from each other and the tags used by one standard are in no way compatible with another due to differences in operating frequency, power harvesting techniques, modulation, and data coding. It is beyond the scope of this lecture to provide a broad description of all the standards, but instead we focus on a few that are currently dominating the industry.

3.1   EPCGLOBAL

EPCglobal is a consortium that created a new *de facto* standard for UHF-based RFID tags. It was originally a grass roots initiative created by MIT's Media Lab. However, due to the success of their Generation-1 tags, ISO are in the process of working with EPCglobal to create a joint Generation-2 standard [21] which contains modifications to enable it to be adopted on a global scale. One of the key additions was the use of a bit in the EPC tag header that differentiates it from the ISO Application Family Identifier, thus enabling RFID readers to distinguish them. At this time it is expected the new ISO 18000-6 standard, which covers the EPCglobal specification will be available by the Fall of 2006.

3.1.1   Generation-1

Within this standard there were two tag classes defined: Class 0 and Class 1.

Class 0: is a read-only identity tag that is programmed during the manufacturing process.
Class 1: is a write-once read many (worm) tag that may be field programmed.

Despite the success of Generation-1 this standard had several limitations. The two tag class definitions did not interoperate with each other and used different wireless protocols. Both classes could coexist, but required two different reader implementations to interrogate collocated tags. This was compounded as RFID manufactures created their own proprietary extensions, such as Matrix (now Symbol) with their class 0++ product. Furthermore, the standard was not ratified as a world standard, and there were many countries that could not use the tag products. In a world of interconnected economies, manufacturers are not likely to adopt such a standard unless they are able to make use of it to sell and ship products within a global economy.

3.1.2   Generation-2

The primary goal for the Generation-2 standard was to create a global standard that would mitigate many of the issues that limited the success of Generation-1. The RF specification is now more flexible and can be used across national boundaries operating in the region of 860–960 MHz and there is a broad support from the majority of technology providers. Furthermore, robustness and read throughput for co-located high-density tag environments has increased. Also to address various privacy concerns the standard now has greater emphasis on secure access control [22]. Fig. 3.2 Shows an early implementation of a Generation-2 tag by Texas Instruments. Impinj Inc. is another important player driving the Generation-2 spec, providing silicon designs ready to be manufactured by the various tag vendors. Four classes of tags are defined that progressively build on the properties of the lower classes. The class properties are listed below:

Class 1 Passive Tags (backscatter)

- write-once read-many to establish a standard EPC identity;
- tag identifier (TID)—information about the manufacturer of the tag (read-only);
- password protected access control;
- kill switch—to disable the tag at POS;
- user memory (optional).

Class 2 Passive Tags (backscatter) Extended Functionality

- re-writable memory;
- extended TID;
- extended user memory;
- authenticated access control;
- additional features—work in progress.

FIGURE 3.1: Format of a 96-bit EPCglobal tag

Class 3 SemiPassive Tags

- an integral power source to supplement captured energy;
- integrated sensing circuitry.

Class-4 Active Tags

- tag-to-tag communications;
- complex protocols;
- ad-hoc networking.

3.1.3   EPC Packet Formats

The EPC tags were defined with the following four fields: header, EPC manager, object class, and serial number (Fig. 3.1).

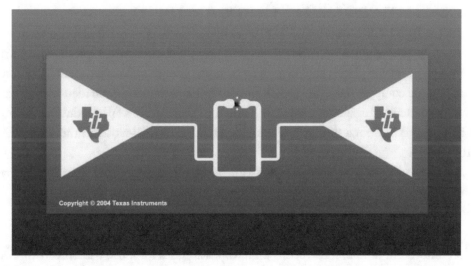

FIGURE 3.2: The Texas Instruments Generation-2 tag (courtesy of Texas Instruments)

Header is an 8-bit field allowing for expansion of the EPC tag format including 64-, 96-, and
    256-bit versions. The most popular of these standards is the 96-bit Universal Identifier
    format. The remaining list describes the fields and sizes for the EPC-96 format.
EPC Manager is a 28-bit field that defines the domain manager for the remaining fields.
Object Class describes the generic type of the product tagged and is 24 bits wide.
Serial Number an individual item number that has 36 bits available to it. The large number space
    provides the first opportunity in the industry for item-level tagging.

Removing the 8-bit header and 28-bit manager bits, there are 60 bits remaining allowing
approximately 12×10^{17} items to be tagged cataloged by each domain manager.

3.2   ISO 15693 VICINITY CARDS AND RFID

ISO 15693 is a standard for both vicinity cards and RFID. Devices can typically operate at
distances of 1–1.5 m and use inductive coupling to provide power and load-modulation to
transmit data. The standard operates in the 13.56-MHz band, and a typical tag must be able
to operate with a magnetic field strength between 0.15 and 5 A/m. For comparison 5 A/m is
about a tenth of the earth's magnetic field strength at the surface.

Tags based on this standard have been widely produced and used for a variety of applications. See Section 5 that describes sports and security applications based on this standard; and
in particular products made by Texas Instruments (Tiris) that support these markets.

A full specification of this standard (four parts) can be obtained from ISO, see http://
www.iso.org/iso/en/prods-services/popstds/identificationcards.html.

3.3   ISO 14443 PROXIMITY CARDS AND RFID

Similar to the ISO 156803 standard, 14443 was created for proximity cards that operate at short
distances. A typical application is fare collection on public transport, typically at a turn-style
requiring a passenger to place a card near to a reader in order to make a payment. As a result the
standard is more complex than ISO15693, providing more than a simple identity and supports
two-way data exchanges. These cards are defined to work with a magnetic field strength of
1.5–7.5 A/m, and thus for a similar reader are designed to be used closer to the reader coil than
with ISO15693.

A full specification of this standard can be obtained from ISO, see: http://www.iso.
org/iso/en/prods-services/popstds/identificationcards.html

3.4   THE NFC FORUM

An important recent development opens up new possibilities for more widespread applications
of RFID. Since 2002, Philips has pioneered an open standard through ECMA International,
resulting in the Near-Field Communication Forum [23] that sets out to integrate active signaling

FIGURE 3.3: The Nokia 3200 cell phone features an NFC reader: Front side—It looks like an ordinary cell phone. Back side—you can see the reader coil molded into the housing (courtesy of Nokia, Inc)

between mobile devices using near-field coupling, and be compatible with existing passive RFID products. The new standard aims to provide a mechanism by which wireless mobile devices can communicate with peer devices in the immediate locality (up to 20 cm), rather than rely on the discovery mechanisms of popular short-range radio standards, such as Bluetooth [24] and WiFi [25], which have unpredictable propagation characteristics and may form associations with devices that are not local at all. NFC aims to streamline the process of discovery by passing MAC address and channel encryption keys between radios through an NFC side-channel, which when limited to 20 cm allows a user to enforce their own physical security. NFC has been deliberately designed to be compatible with ISO15693 RFID tags operating in the 13.56 MHz spectrum, and allow mobile devices to read this already popular tag standard. It is further compatible with the FeliCa and Mifare smart card standards that are already widely used in Japan.

In 2004, Nokia announced the 3200 GSM cell phone (Fig. 3.3) that would incorporate an NFC reader. Although the company has not published an extensive list of the potential applications, it can be used to make electronic payments (similar to a Smart Card) and place phone calls based on the RFID tags that it encounters. For example, at a taxi-stand a prospective client might bring their phone near to an RFID tag attached to a sign at the front of the waiting area. The result would be a phone call to the taxi company's coordinator and a request for a car to be sent to that location [26]. This model allows a close link between the virtual components of our computer infrastructure and the physical world, such as signs and taxis, and is a key enabling technology that contributes to the implementation of the Ubiquitous and Pervasive Computing vision as proposed by Mark Weiser [27].

A complication for the wide-scale adoption of the NFC standard is that state-of-the-art EPCglobal RFID tags are based on far-field communication techniques, working at UHF frequencies. Unfortunately, NFC and EPCglobal standards are fundamentally incompatible.

CHAPTER 4

Reading Collocated RFID Tags

One of the ultimate commercial objectives of RFID systems is the ability to read, and charge for, all of the tagged goods in a standard supermarket shopping cart by simply pushing the cart through an instrumented aisle. Such a system would speed the progress of customers through checkout areas and reduce operational costs. The solution to this problem can be thought of as the holy grail of RFID technology. It has many engineering issues that make it difficult. First, the RF environment inside a shopping cart is particularly challenging. The product packaging in the cart is made of a wide variety of materials that include metal cans and aluminized plastics that reflect and shield the interrogation signals. Furthermore, some of the products contain water, and plastics, that may absorb RF signals in the microwave band. To complicate matters further, all of the products are in close proximity to each other and in random configurations. RFID tags attached to these products will sometimes be poorly orientated with respect to the reader's antenna, thereby making RF communication difficult. In addition, tag antennas are typically flat to enable them to be embedded in labels, but if orientated edge-on to the reader the tag will likely not receive enough energy to power up. These specific problems are discussed in more detail in Section 10; however, even if the RF reading environment for a group of RFID tags is ideal, it is still an engineering challenge to design readers that can successfully query multiple collocated tags, and accurately determine all of their IDs in a short period of time.

Consider two tags that are situated next to each other and equidistant from the reader. On hearing the reader's signal they will both acquire enough energy to turn on and then transmit their response back to the reader at roughly the same instance in time. The result will be a collision between the two signals, and the data from both tags will be superimposed and garbled as a result. Collisions can be detected at the reader's receiver by augmenting its demodulation circuit to look for signal encodings that contain an anomalous format. For multinode communications networks, such as Ethernet, this is a well-understood problem addressed in protocols such as CSMA/CD [28], or 802.11 that uses a variant of MACA (Multiple Access/Collision Avoidance) [29]. The solutions employed are based on arbitration protocols providing the colliding nodes with a new opportunity to successfully deliver their data while minimising the resulting wasted channel bandwidth.

Arbitration using a statistical approach has been implemented in some RFID systems by inserting a random delay between the start of the interrogation signal and the response from the tag. But even if each tag randomizes its response time there is still a finite probability that a collision will occur, and the reader must carry out several rounds of interrogation until all of the tags in that area have been heard with high probability. This algorithm can be enhanced further by using a protocol that prevents tags that have already been heard from responding again until the current interrogation cycle has ended. At each interrogation request there will now be a progressively smaller population of tags that will respond, reducing the likelihood the remaining tag responses will collide.

Unlike a general purpose wireless communication network in which an ad-hoc collection of nodes have equal status, an RFID reader has a privileged position and can centrally orchestrate an arbitration protocol. This allows for a more deterministic arbitration protocol to be used. The process of uniquely determining a tag's ID from the surrounding population of tags is sometimes called *singulation*.

4.1   QUERY TREE PROTOCOL

EPCglobal Generation-1 class-0 tags use a Query Tree Protocol to singulate tags. Figure 4.1 shows an example of how three unique, but collocated tags (*001*, *100*, and *110*), can be successfully read using this protocol. The reader R starts an interrogation (level-0) by asking which

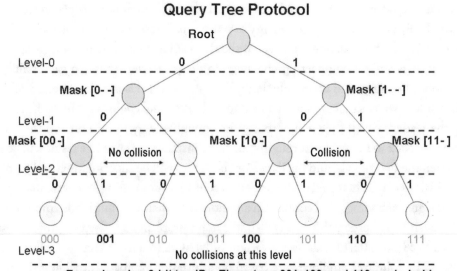

FIGURE 4.1: Arbitration mechanism used by EPCglobal Generation-1 Class-0

of the top branches of the tag identity-space (modeled as a binary tree) contain tags. It does this by broadcasting a prefix that initially selects the left branch in the tree; in our example the bit-mask would be [0–] selecting all tags with codes 0XX, where X is either a 0 or 1. A mask operation is achieved by only considering the number of bits in the prefix and logically XORing the mask value with the corresponding top-bits in the Tag's ID, and then looking for a zero response to determine a match. If a tag matches the condition it will respond with the value of the bit that follows the prefix (either '0' or '1').

If only one tag exists in the left sub-tree, as in our example *001*, its next bit '0' will be returned to the reader *R*, and it can make an additional query with prefix [00–] to find the final bit of the sequence, thus singulating tag *001*.

If multiple tags are present matching this condition, as in the right branch example detected by prefix [1–], they will all respond and a collision will result. A collision situation is recorded at the reader as aberrations in the received waveform. As a result the reader *R* will respond by separately querying each of the sub-trees below this point in the tree by separately transmitting the query prefix [10–] and [11–]. Each will result in a tag response without a collision, and thus singulate the ID for *100* and *110*.

Although the process above may seem complicated, the same steps are applied repeatedly at each of the sub-trees in depth-first-search order, and by using recursion to implement the Query Tree Algorithm; the process can be defined concisely, even for an arbitrary number of bits in the tags. An important aspect of the algorithm is that when there is no response from one of the sub-trees, it is removed from the tag search-space. Thus, the queries need only be applied to parts of the tree that contain tags, and after a short time all tags present will be able to respond to the reader in depth-first-search order. The cost of the algorithm is bounded by the number of bits in a tag ID n times the number of tags being read. In practice, tags are likely to be allocated to organizations in sequential batches that will tend to localize the tag identities in particular sub-branches of the code space. This will accelerate the search process, because the amount of recursion needed to complete the algorithm will be reduced.

Manufacturers of EPCglobal Generation-1 based their arbitration mechanism on this algorithm, and claim it is possible to accurately read up to 500 collocated tags per second. One of the advantages of this arbitration algorithm is that it does not require any state to be held in a tag itself, instead a reader has the responsibility of probing the tag identity space, managing the recursive queries, and keeping track of the branches of the tree that contain tags. However, this approach does raise a privacy concern. As the reader homes in on the tags it identifies, it broadcasts their IDs using the full power of the reader's transmitter. This means that a distant eavesdropper with a suitable receiver may be able to record the IDs being scanned. And more of a concern, the products being purchased can also be identified. Many of the undesirable consequences of unwittingly disseminating this type of information are discussed in Section 8 on *Privacy*.

In order to overcome this limitation, the EPC Generation-2 arbitration protocol was designed to avoid the reader transmitting the tag IDs at high power. The essence of this algorithm is described in Section 4.2 below.

4.2   QUERY SLOT PROTOCOL

Query Slot Protocol is an arbitration mechanism used as the basis for singulation in the EPC-global Generation-2 tags. Unlike the Query Tree Protocol Algorithm described in 4.1, it has an advantage that the reader does not need to transmit the IDs of the tags in order to determine the inventory. In this algorithm only the tags backscatter their ID thus limiting the range at which the IDs can be detected as radio signals. However, it requires additional state registers available in each tag, and increases the complexity of the design. But yearly progress in the capabilities of CMOS integration, as predicted by Moore's law, allow for considerably more transistors to be fabricated in the same area of silicon. Today, the additional complexity of adding a few data registers, and the associated protocol state-machine, does not add a significant cost burden to the manufacture of an RFID tag.

An example of the Query Slot Protocol is shown in Fig. 4.2. In the specification there are other aspects to the protocol, but this simplified description helps us understand the core mechanism. To make a comparison with the Query Tree Protocol shown in Fig. 4.1, the same three tags with IDs *001*, *100*, and *110*, are used as an example collocated tag inventory that must be successfully read.

The algorithm requires tags to provide the following capabilities. Each tag contains a counter *count*, initially set to zero; an inventoried flag; initially cleared, and a random number generator that can produces 16-bit values *rn16*.

The reader *R* transmits a *QueryRequest* command to the tags with a parameter *Q*. All tags that hear the command start an inventory round and clear their inventory flag, they also enter an *arbitrate* state. They then generate a *Q*-bit random (0 to $2^Q - 1$) and load this into their slot counter *count*. If a tag's count is zero, as it the case of tag *001* in the example, it will be in a *reply* state and will generate a 16-bit random number *rn16* and backscatter it to the reader. The reader will then respond with an acknowledgement including the *rn16*, and if a match is found, will backscatter its *EPC* code, which can be recorded by the reader and stored in an inventory database. This tag will now set its inventoried flag and go to sleep until it hears a new *QueryRequest*.

In order to find all of the tags in the inventory, the reader will now initiate a *QueryRep* (repeat) command which will cause all remaining tags to decrement their slot counter, *count* = *count-1* and once again, any tag with a value of zero will respond with an *rn16* value and the

Generation 2: Query Slot Protocol Example for a Reader R and Tags T()

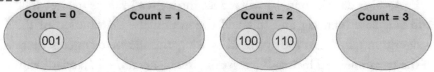

Tags

1. **Reader R:** Sends query request with parameter: Q (Example Q = 2) and initiates an inventory round.

2. **Tags T():** Load an internal slot counter with a random Q-bit number and clears inventoried flag.

SLOTS

Count = 0	Count = 1	Count = 2	Count = 3
001		100 110	

Example Q = 2, resulting in four slots, RN16 is a 16-bit number

3. **Tag with count=0 (e.g.,001) backscatters an RN16 random number.**

4. **Reader R:** Acknowledges RN16 number.

5. **Tag (e.g.,001) checks RN16 matches and backscatters EPC ID.**

6. **Reader R:** Issues QueryRep command
    Tag 001 set Inventoried Flag, and goes to sleep
    Tags T() remaining decrement slot count
    Loop to 3 until 2^Q-1 QueryRep commands

FIGURE 4.2: Example of the Query Slot Protocol used in EPCglobal Generation-2

process will continue as before. If by chance each tag had found a unique slot, after $2^Q - 1$ *QueryRep*s had been issued, all of the tags would have been inventoried.

However, in our example there are two tags that have the same *count* value, *100* and *110*. When their count is zero both tags will be in the *reply* state, and respond at the same time with different *rn16* numbers and a collision will result. This is detected by aberrations in the signals detected at the reader and it will instead respond with a *nack* and a *QueryAdjust* with a new *Q*-parameter without changing the state of any of the inventoried flags in the tags. The remaining tags will again randomize their slot counter value *count* with a $2^Q - 1$ number and the process will continue until all tags have been singulated.

An interesting aspect of this protocol is that the optimal value of *Q* is dependent on the number of tags being inventoried. Ideally *Q* should be chosen so that there will be as many slots as there are tags. If *Q* is too large there will be too many slots that are empty for each *QueryRep*, and if too small there will be too many collisions. However, the number of tags is initially not known, and therefore the value of *Q* must be found experimentally by testing the environment and increasing or decreasing *Q* until an acceptable arbitration behavior is found.

4.3   SUMMARY TAG READ RATE TIMING FOR EPC GENERATION 1 AND 2

To provide some context for the tag read rates used in the algorithms described above, it is useful to consider timings used by the EPCglobal Standard. In the EPC Generation-1 tag the reader-to-tag command timing is approximately 1 ms, and the tag reply 0.9 ms. The total round trip of ∼2 ms leads to the claim of a maximum read rate of 500 tags-per-second. However, a Query Tree Protocol as described in 4.1 will reduce the tag read-rate as the tree walk procedure takes time to singulate each tag and this will depend on the distribution of the IDs. The binary tree algorithm minimizes the overhead by a logarithmic factor rapidly descending into the parts of the tree populated by tags. Comparing the Generation-1 standard with the future Generation-2 standard, Generation-1 systems read tag-data at 140 kbps, whereas Generation-2 has an adaptive scheme from 40 to 640 kbps. Generation-2 EPC-96 bit IDs can potentially be read over four times the speed of Generation-1 tags. The Generation-2 Query Slot Algorithm described in 4.2 can thus have arbitration slot times as small as 0.5 mS.

CHAPTER 5

Applications of RFID Tagging

In this section we examine the wide variety of applications that can take advantage of RFID. The capability of identification at a distance can be extended to include sensing applications and read/write memory, however, these topics will be covered later in Sections 6 and 9 respectively as they warrant special discussion. Below we provide a categorization of mainstream RFID applications:

- Security:
  ○ access control: keys and immobilizers;
  ○ patrol verification: process management;
  ○ antitheft: merchandize.
- Tracking:
  ○ supply chain: warehousing and inventory control;
  ○ people and animals: personel, children, patients, runners, cattle, and pets;
  ○ assets: airline luggage, equipment, and cargo.
- Authenticity:
  ○ money: banknotes;
  ○ pharmaceuticals: packaged drugs;
- Electronic payments:
  ○ transportation: auto-tolls: FasTrak, EZ-pass;
  ○ ticketing: ski passes;
  ○ credit/debit cards: PayPass by MasterCard.
- Entertainment:
  ○ smart toys: interactive characters.

5.1 SECURITY

It is possible to divide RFID-based security applications into three subcategories: Access Control, Verification, and Antitheft.

5.1.1   Access Control

One of the first applications that motivated the design of modern passive RFID was Access Control, enabling mechanical keys to be replaced by an electronic card. The primary advantage is that card keys are harder to forge and it is much easier to revoke a key that has been compromised or lost, simply by deleting it from the access database, or to issue a security alert if a revoked key is used nefariously.

This kind of RF tag has been improved since its early beginnings in the 1960's and is now widely adopted as the basis for corporate identity badges by numerous organizations around the world. In addition to verification of an employee's identity, they often serve as proximity cards providing access to corporate campuses, buildings, and laboratories; and reduce the need for security guards at all of the entrances. The read range is usually limited to a foot or less to avoid unintentionally opening a door, but still has the advantage over a key that the badge can be left in a bag or a wallet, and provide convenient access for the owner without needing to physically remove it.

In comparison to inventory control, RFID-based access control is not a cost-sensitive application. ID cards have a long life and may incorporate other prerequisite features necessary even without RFID, such as a photograph and a robust laminated plastic form factor. One of the manufacturers of these cards is Hughes Identification Devices (HID) that provide a wide range of RFID solutions. Their contactless ID cards operate either at 125 kHz or 13.56 MHz (depending on local spectrum legislation) and can also store between 2 and 16 kbits of read/write data (Fig. 5.1).

Another, now common, form of RFID key is used in automobiles to make it harder for vehicular theft. An example is the Chrysler Jeep that incorporates an RFID tag into the body of the ignition key. The lack of the correct RFID tag serves as an immobilizer. The car will only start if both the unique mechanical key and the unique RFID tag are present. Thus, even if a would-be thief were able to make an impression of the key and covertly reproduce it, the forgery would be of no use without the embedded RFID tag. And similarly a smart RFID reader that can determine the tag's identity and later masquerade as the tag, would also fall short as a key. The requirement that both a unique RF interaction and a physical key are present considerably raises the technical skill necessary for a successful forgery (Fig. 5.2).

5.1.2   Verification

Most companies and government institutions employ security personnel to guard their entrances, and make periodic checks that their campuses are secure. Despite modern electronic measures to help with this task, such as the use of security cameras, the most versatile line-of-defense is to employ a security patrol to periodically make "the rounds" and check for suspicious activity. Employers may find themselves legally bound to ensure that security measures are being undertaken, and an insurance company might also require verification of these measures before

FIGURE 5.1: The iClass RFID identity card from HID

providing coverage. A traditional method for verification of a security patrol has been the use of log stations; a guard is required to punch a card at a log station validating the location and time of the patrol. A modern alternative requires that guards carry a handheld RFID scanner which electronically logs the time strategically placed tags in the building are read. The scanner can be interrogated at the end of a work shift to ensure the guard was present at all the critical

FIGURE 5.2: Automobile ignition key with additional RFID activation

FIGURE 5.3: RFID tags and handheld readers assure security patrols are carried out consistently and according to a predefined schedule (courtesy of Proxiguard)

locations and at the required times. This idea can be extended beyond security to any task that needs periodic verification for location-based activities (Fig. 5.3).

5.1.3 Antitheft

Automatic mechanisms to protect a store's merchandize from shop-lifting have been used for many years. The common antitheft tag is a simple device that is attached to merchandise in a store and disabled at the check-out desk at the time of purchase. However, in the case of theft, the tag will not be disabled and trigger an alarm at the exit of the store. This kind of tag is usually based on an inexpensive resonate circuit that can communicate its presence through load modulation (see Section 2—Principles of RFID). However, these tags can be thought of as binary RFID tags, indicating their presence when interrogated, but do not provide any additional information beyond this. However, if RFID tags become common for item level inventory control, they can also serve a dual purpose providing an integrated antitheft capability, and may eventually replace the simple binary tags.

5.2   TRACKING

There are numerous examples of large organizations that need to track the location of equipment or people in order to operate efficiently. This is a logistics problem that on a small scale is easily handled by well-trained people, but on a large scale can only be achieved effectively by automation. For example, consider the problem of tracking goods that might be waiting for shipment from a factory, or in transit, or have just arrived at a distributor, or on the shelf of a retail store in one of many possible locations. These problems can be mitigated by the use of automatic identification, computer networks, and computer databases, which can be rapidly queried and searched to provide answers. Large retailers, health organizations, and military operations, can all benefit from automatic tracking enabled by RFID.

5.2.1   Supply Chain and Inventory Control

A driving force behind the widespread adoption of UHF based RFID is supply chain management. The potential for lowering the operational cost of the supply chain is what motivates Wal-Mart, Tesco, Target, and other major retail stores to adopt UHF RFID into their work practices. Even in a warehouse, inventory can be lost or misplaced, and RFID systems are well suited to finding its location. This is due to a tag's long read range, resilience, and the property it can be read without requiring line of sight. The latter property has the corollary that it is possible to continuously scan for RFID tags, and the supporting systems can therefore continuously track the comings and goings of inventory without human intervention. RFID systems are well suited to mitigating human error in warehouse environments which are dynamic and sometimes hectic environments, resulting in a more efficient operation. More details about the Wal-Mart RFID trials are presented in Section 7.

5.2.2   People

Sub-dermal tagging of animals is more socially acceptable than sub-dermal tagging of people. The suggestion of tagged people immediately brings up images of George Orwell's novel "1984". However, there are already examples of injectable RFID tags that have been applied to people. Kevin Warwick, and professor of Cybernetics, at the University of Reading experimented with RFID in 2000. Placing a tag under the skin of his own arm and using it as a unique key to gain access to his house (Fig. 5.4). He described his experiences in Wired Magazine [30]. In the late 90s, Applied Digital Solutions produced the Veritag design specifically for tagging people

FIGURE 5.4: Subdermal RFID tag in the arm Kevin Warwick (University of Reading) (*Wired*, February 2000)

using a sub-dermal glass capsule; other human injectable tags are also made by Trovan Ltd., in the UK. The technology has found a number of niche applications. In some parts of the world such as Mexico, kidnapping is more prevalent than in the US, and tagging children may help parents identify them years later, and possibly also serve as a deterrent.

5.2.3   Hospital Patients

RFID has application in the health care industry for tagging patients to ensure that medical records are correctly associated with the people they describe, and that the correct medications are administered. These records can also provide information about a patient's allergies, and is therefore critical for this association to be made correctly. Printed labels, or even typing in a name on a computer keyboard in plain text, can lead to simple mistakes, e.g., Mrs. I. Smith versus Mrs. L. Smith. However, many hospitals currently use barcoded wrist bands which also solves the problem, and it is unclear if RFID will improve this work practice.

5.2.4   Runners

Since the late 90s the organizers of major marathon races in the US have provided runners with an RFID tag that can be incorporated into the laces of their running shoes (see Fig. 5.5). This became necessary to handle the logistics of timing 1000s of runners in a major metropolitan race, and a solution to the problem of managing so many staggered start times because it may take at least an hour for so many runners to cross the start line and begin the race. The system operates by employing RFID readers at the start and finish, and other key checkpoints along the race course. As the runners pass the tag reader stations, a time is recorded for each ID and

(a)   (b)

FIGURE 5.5: (a) An RFID tag on a shoelace (courtesy of Texas Instruments). (b) This system is used in the Boston Marathon

(a)

(b)

FIGURE 5.6: (a) RFID tag mounted in the ear of a cow. (b) Dog with subdermal tag being identified with a handheld reader

thus at the finish, the running time for each contestant can be calculated automatically as they complete the course.

Furthermore, with several checkpoints on route it is possible for the organizers to provide breaking news of how the top runners are performing while the race is in progress. This information is also of great interest to the athletes after a race, because the split times help them understand how they performed through the event, and at what stage in the race they might be able to push themselves in the future.

5.2.5   Cattle

Managing a modern dairy farm requires detailed accounting for the entire herd. This includes monitoring how much they eat and a list of all the medications that have been administered. Keeping track of the identity of each animal is therefore important, but conventional labels such as barcodes cannot be used in the dirty environment of a farm. RFID is better suited as it is not affected by soiling. Figure 5.6a shows an ear tag that has been designed for use under these conditions. In recent years there has been cause for concern that livestock may contract Mad-cow disease, and as a result there is more interest than ever in tracking the ownership and medical history of a cow throughout its lifetime.

5.2.6   Pets

Many States in the US provide a service for owners to have their dogs and cats electronically tagged using an injected RFID tag behind the ear or neck. If lost and then later found, authorities recovering the animal can scan for a tag to identify the owner's name and address. Stray animals often lose their collars, and thus sub-dermal RFID is more likely to survive the ordeal. Figure 5.6b shows a commercial reader designed for this purpose. Unfortunately, there is no US-wide standard for the type of tag used for this application, and States use tags that are not compatible with the result that some tags are not detected. This is an example of why adopting uniform standards is so important for successful RFID applications.

FIGURE 5.7: An airline luggage label that provides RFID, barcodes, and printed information about the owner, flight, and destination of the tagged bag (courtesy of Texas Instruments)

5.2.7   Airline Luggage

In an era of heightened terrorism concerns it is important that airline baggage be tracked and ideally travel on the same plane as its owner. If bags can be automatically identified while being moved to a plane we can have greater confidence the process will be error free. Also, if a passenger has checked in but later does not board the plane, RFID can help locate the bags in the cargo hold for removal before the plane takes off. Figure 5.7 shows how traditional labels can be integrated with RFID allowing backward compatibility with existing barcode systems.

A further benefit is to help mitigate the costs associated with recovering a passenger's lost luggage and delivering it to the correct address. The US Bureau of Transportation claimed that over a billion items of checked luggage were transported in 2004 and if only 0.5% were misrouted, this represents 5 million lost bags. The accumulated costs of these mistakes can be a significant burden for an airline's operating costs, and automated RFID tracking can help remove the human errors that lead to this problem.

5.3   AUTHENTICITY

In order to have confidence that any manufactured item of value comes from an authentic source, it is necessary to have a means of validating its origin. Examples of authentication mechanisms from other industries include hallmarks for items made from precious metals, watermarks on banknotes, and artist signatures on paintings. However, all of these can be forged if investment is made in the appropriate equipment, but provide a high enough financial hurdle that in practice a forgery is a relatively rare event.

5.3.1 Money

The Mu-chip from Hitachi is one of the smallest RFID implementations at 0.4 mm × 0.4 mm and designed to be read at a very close range. The EEC have been considering embedding the Mu-chip in future ECU banknotes, primarily to provide an automatic means of validating their authenticity, and for rapidly counting a pile of notes. Such automation removes the chance of human error when reading denominations, or when separating notes that may have become stuck together. However, adding a unique number that is automatically readable detracts from paper money's most valuable asset—it is not easily traceable. Tagged banknotes may be automatically tracked between transactions and thus provide information about when and where you spend your money. In contrast, credit cards have long since given away this information, but the use of cash in transactions has preserved our privacy. Adding RFID tags to banknotes is potentially another area that may erode our fundamental right to privacy (Fig. 5.8).

5.3.2 Drugs

Pharmaceuticals often have a high market-value and are therefore a target for forgery. However, unlike many consumer products, it is difficult for the lay person to know if the pills purchased in a bottle are really the drugs they claim to be. We usually rely on the reputation of the dispensing pharmacist to validate the purchase, but in the era of the Internet there are often attractive drug purchases to be made online. Off-shore companies may well have lower operating costs and can legitimately provide bargain drugs, but a consumer can no longer be certain they are getting

FIGURE 5.8: Hitachi's Mu-chip—so small it can be sandwiched between the paper layers of a banknote (courtesy of Hitachi, Ltd)

(a)

(b)

FIGURE 5.9: FasTrak Toll Pass System: (a) Transponder tag and (b) Booth

what they paid for. Sealed packaging that includes hard-to-forge RFID tags with batch numbers allocated from the pharmaceutical manufacturer are a solution. The batch number could then be read at home and validated online through the Internet. See Section 7 and antitamper proof packaging.

5.4   ELECTRONIC PAYMENTS

5.4.1   Auto Tolls

For many new road and bridge projects, the use of tolls has been the only way to raise the capital investment to pay for them. However, tolls are inconvenient for drivers, who need to carry the appropriate change. Traffic is also slowed, often leading to traffic jams at peak commuting times. RFID technology can reduce these problems.

By placing a suitably designed tag in the windshield of a car (see Fig. 5.9a) a tag reader at the tollbooth can automatically scan its ID as it passes by. The systems are designed so that customers establish a prepaid account and a booth can then deduct the appropriate fee from the account each time the ID is detected. The toll-reader technology has been developed so that it can operate at freeway speeds and thus cars, in theory, do not need to slow down. However, in practice most tollbooths are narrow (see Fig. 5.9b), and cars are required to slow down for safety reasons, but even at speeds of 20 mph there is a significant increase in traffic throughput.

In recent years more lanes at toll plazas have been converted to provide automatic toll charges. On the west coast of the US, FastTrak is the predominant system, whereas on the east coast, EZ-pass is more common.

5.4.2   Electronic Tickets

Ticketing is another domain where RFID can provide a unique contribution. A ticket is a prepaid token that provides personal access to a resource, for example, a movie, exhibition, or

FIGURE 5.10: Ski pass with embedded RFID serves as a ticket to enter the chair lift

museum. Usually a ticket must be purchased ahead of time, but when it is used, no additional money is required. Furthermore, a ticket can only be used once (although multiple tickets can be held in one physical token) and they usually have a limited lifetime. The main purpose of a ticket is to provide rapid access to an event when large numbers of people are converging on the same location. Any delays in providing payment, giving change, or validating that credit is available, are decoupled from the process of gaining access. Access to the event is achieved by simply surrendering a valid ticket.

An RFID-based ticket has greater advantage over a traditional paper counterpart as it can be left in a pocket while being validated, and electronically stamped as 'used', when the patron passes through a turn-style. It can also store several virtual tickets in the same device and surrender each one as required; or a single token can enable access for a limited period of time. An RFID-based ticket can also be renewed electronically and thus can be used on multiple occasions, which can offset the cost of the technology in comparison to printing tickets. RFID has already been used to implement high-value tickets such as those found at a ski resort to gain access to chair lifts (Fig. 5.10), and to provide daily access to the subway in Tokyo using the Milfare system.

5.4.3   Electronic Credit

PayPass is a payment token being pioneered by MasterCard to provide a fast and convenient method of buying low-value items. The PayPass token can be in the form of a card or a key fob. To buy merchandise, a customer can use the token based on "contactless" RFID technology (ISO 14443) to make a payment, simply by moving the token in front of a reader. No signature is required for items below $25, making the transactions fast and convenient for small purchases, but for items of greater value a PIN or signature is required. PayPass is expected to extend and

FIGURE 5.11: A Star-wars character from Hasbro. Placing different characters on the podium plays the sound and voice of that character

augment the current network of magnetic strip readers available for MasterCard purchases and has been in trials since 2003.

5.5   ENTERTAINMENT

5.5.1   Smart Toys

The invisible nature of RFID communication has been used by some toy manufacturers to create toys that appear to magically take on a personality when brought near other objects. The Hasbro Star Wars character (Fig. 5.11) contains an RFID tag and when placed on a podium, generates sound effects and speech associated with that character. Although the toy could have been designed so that its tag had a simple unique ID that triggered the podium to play a corresponding audio file, the designers felt is was more flexible if the RFID tag stored the entire audio clip and the reader simply had to play the file that it was able to read. Using this approach, new characters could be sold without updating the software in the podium, providing greater flexibility for the manufacturer, and at the same time, keeping the consumer experience very simple.

CHAPTER 6

RFID Incorporating Sensing

One of the most intriguing aspects of modern RFID tags is that they are able to convey information that extends beyond an ID stored in an internal memory, and dynamically read data from an on-board sensor [31]. Today, there are commercial implementations of RFID technology that can verify that critical environmental parameters remain within a safe range, and as a result can be used to ensure the integrity of perishable goods, and protect the interests of retailers and customers alike.

In the following sections we look at the various categories of RFID sensor in detail. Specifically, we examine the characteristics of sensing applications that are suitable for integration with RFID technologies, and provide examples of monitoring physical parameters such as temperature, pressure, and acceleration; tamper detection; chemical or bio-agent detection; non-invasive medical monitoring; the use of memory in combination with sensing, and techniques for building longer range sensors.

6.1 EXTENDING RFID TO SENSING APPLICATIONS

The same mechanisms that enable an ID to be read from an internal register in an electronic tag can also be applied to collecting data derived from a sensor. Extending the capabilities of the silicon chip to interface with a sensor is straightforward, but the design of a suitable sensor is usually an engineering challenge. First, the sensor will not be able to scavenge energy from the reader while the tag is out of range; and this is likely to be the predomiant state during the tag's lifetime. Second, even when it is in read range, the available energy is very small. As a result, this limits the capabilities of the electronics that can be used to process and record a sensor reading.

6.2 MONITORING PHYSICAL PARAMETERS

An important application of RFID sensing is in the realm of monitoring perishable goods. Typically items such as meat, fruit, and dairy products, should not exceed a critical temperature during transportation, or they may not be safe for consumption at their destination. An RFID temperature sensor can serve to both identify and track crates of perishable goods, and ensure their critical temperature has remained within recommended parameters [32,33]. An example is the KSW TempSens RFID tag [34] (Fig. 6.1) which has been designed explicitly

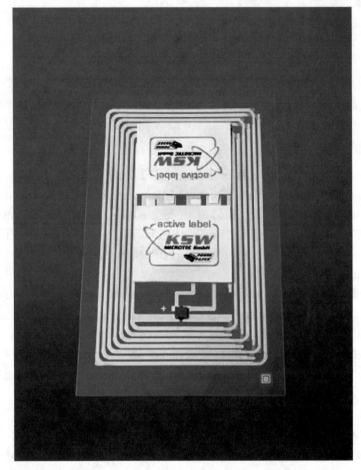

FIGURE 6.1: An RFID tag that can detect a critical temperature threshold (courtesy KSW Microtec AG)

for this purpose, can be integrated with a standard product label. An example application is illustrated by the challenges related to the transport of frozen chicken, which has a high risk of salmonella contamination if allowed to thaw (Fig. 6.2). Furthermore, if later frozen again, it may not be apparent from a visual inspection that a problem had occurred. A temperature monitoring tag operates by incorporating a material in the tag's substrate that makes a permanent electrical change when the critical temperature has been exceeded. This can be represented as a single binary digit appended to the ID, or if more bits are available, a measure of the maximum temperature exposure. When the tag is read, it will not only respond with an ID, but also provide a warning if the temperature variation has been an issue.

Monitoring the pressure of an automobile's tires from inside the vehicle is another application that is well suited to the unique capabilities of remote RFID sensing. This is a feature that can be of benefit to drivers, as a slow leak often goes by unnoticed before the tire becomes completely flat. By the time the fault is discovered, tire damage and considerable inconvenience may result. This is a challenging problem because the tires are spinning and sensors can only be connected by wires if concentric connection rings are engineered around each wheel axial, making the engineering complicated and costly. Furthermore, even a wired connection to an external sensor placed on the opening of a tire valve is likely to be unreliable, and subject to damage depending on how the car is driven. Instead, some companies such as as Royal Philips Electronics have been developing an RFID chip that can be bonded inside a tire, and read remotely by antennas installed in the wheel hubs of the car. As there is no physical link, reliable interaction with the tag is possible even while in motion. The tags can also provide additional information about the tire's maximum inflation pressure and identify the tire for record keeping, e.g., front–back rotation history. Currently, Michelin and other major tire vendors are trialing RFID pressure monitoring systems, but they have not as yet made it to the mainstream automobile market (Fig. 6.3).

Another physical parameter that can be monitored to useful effect is "acceleration." If a fragile package has been dropped during transport, it is likely a critical acceleration threshold would have been exceeded. Today, some shipping companies employ a nonelectronic tag solution to solve this problem, which utilizes a thin plastic membrane to hold a colored dye. If the membrane breaks after an impact, the dye flows into a visible chamber and the tag changes color. This kind of tag is used to detect poor handling in a warehouse, or loading and

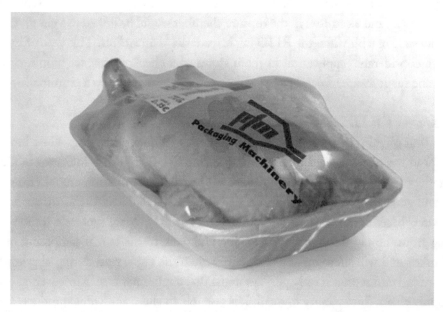

FIGURE 6.2: A packaged chicken incorporating an RFID temperature sensor (courtesy KSW Microtec AG)

FIGURE 6.3: Car tire incorporating RFID pressure sensor readable from the car

unloading mishaps during transportation. RFID adds more utility to this solution by enabling the automatic detection of a damaged item without having to inspect each package by hand. To incorporate this capability into a tag, the dye in the example above can be replaced with a conductive liquid, and electrodes in the rupture chamber would be linked to a circuit that in turn changes the state of a bit when the RFID tag is read; the 'damage' bit. This type of information could have considerable application in retail stores as a checkout cashier monitoring the tag reader can be made aware of damage before passing the product on to a customer.

6.3   TAMPER DETECTION

RFID sensing can also be used to support antitamper product packaging. Most modern consumable products are protected by packaging technology that clearly indicates to a customer if the product has been tampered with. Devices that detect tampering are relatively straightforward and generally require a simple, single-bit interface to detect the alarm state. A simple binary switch-based sensor can be incorporated into an RFID tag, such as a thin loop of wire extending from the tag through the packaging and back to the tag. If tampering occurs, the wire is broken and will show-up as a tamper bit when the tag is read during checkout. In this way, a store can ensure that it only purveys items that are tamper free. Moreover, at each point in the supply chain, from factory to retail, it is possible to check individual products for tamper activity making it easier to find where a crime has been carried out.

6.4   DETECTING HARMFUL AGENTS

In today's political climate people and governments worry that chemical, biological, or radioactive agents might be used by terrorists to threaten populated areas. Many of these agents are themselves invisible, which adds to the danger because people might be unknowingly exposed for long periods of time. The risks can be minimized if it is possible to build detectors that can rapidly alert a population to the presence of the contaminant. Furthermore, if these detectors could be deployed at a country's ports of entry, it might help to identify terrorist plots before they can be put into effect.

A problem with many conventional contaminant detectors is that they are relatively expensive and cannot be deployed on the scale necessary to effectively protect a metropolis. An RFID sensor based on passive detector technologies can be deployed more ubiquitously. The reader part of the system, which is more expensive, can then be installed on vehicles or carried by security personnel. The readers, configured to automatically interrogate nearby tags, would provide a warning about the contaminant as they passed by. At present, sensors that detect biological agents are very limited in scope. A great deal of work needs to be done in this area to build passive detectors that are both effective and inexpensive. However, RFID can be used as the reporting mechanism to make this kind of sensor practical.

A further example is detecting bacterial contamination of food products through routine handing. Although some problems can be detected indirectly using a temperature sensor, a more direct indication is given by testing a sample for bacteria growth. Auburn University's Detection and Food Safety department is carrying out research that will allow them to build an RFID tag providing a measurement of the growth of a particular organism (Fig. 6.4).

6.5   NON-INVASIVE MONITORING

Advanced medical monitoring can also be supported by RFID. Some diagnoses can only be made when there is direct access to the internal organs of the body—even advanced MRI scanning has its limitations. Advances in biopsy techniques and keyhole surgery provide a partial solution, but

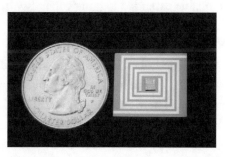

FIGURE 6.4: Auburn University RFID bacterial sensor chip (www.auburn.edu/audfs) (courtesy Auburn University)

some conditions are progressive and call for continuous monitoring without repeated surgery. This is where an RFID tag can play a valuable role. RFID sensors can be designed to be placed in the human body during surgery. An external reader can then be used to periodically communicate with the device either during routine visits to the doctor, or as a result of being carried by the patient. A device such as this can provide an on-going and progressive evaluation of the condition being monitored. Such designs are only in their infancy, but implantable temperature sensing devices are currently being designed by Silicon Craft Technology [33], allowing an accurate body temperature reading to be obtained for livestock implanted with the device. As a result it is easier to detect infection and take the appropriate action early on. Such devices may also be used with poultry to detect the onset of the deadly Avian Flu.

6.6   LOGGING SENSOR ACTIVITY

Detection of an unusual sensor reading helps us know if a physical trigger has occurred, but it would be even more useful to know where and when it happened. Unfortunately, without a conventional battery, RFID sensors cannot continuously monitor the state of a sensor or utilize an electronic clock and automatically record the time of a sensor event. However, readers with accurate clocks (see Section 9) can help in this process by utilizing a tag's on-board writable memory, and recording read-time and sensor-state in a circular buffer. As a result, the time of an aberrant sensor reading can be bounded by the prior and subsequent readings in the buffer.

The value of sensor data can be increase futher by adding location information. A reader equipped with a Global Positioning System (GPS) can write the reader location into the tag, along with the current time and sensor state. If GPS is unavailable, a low cost alternative is a location reference tag near by that can be read to initialize the location of the reader before it scans other nearby tags. The electronic memory in a population of tags can thus serve as a distributed database of the sensing history, including time and location, without requiring that all the readers coordinate their scanning activity through an external network. Although such systems are not available today, the growing use of RFID tags, and the availability of on-board writable memory, is likely to enable extended logging capability in the near future.

6.7   LONGER RANGE SENSING

Most of the applications for sensing described in this section have been based around short-range RFID tags that incorporate sensing, typically using inductive coupling to derive their energy, and load modulation for data return. There are many other categories of sensing application that need to communicate over much greater distances. In recent years some of these applications have been a research topic addressed by the ad-hoc sensor-network community [35]. Remote sensing is achieved by building a wireless multi-hop ad-hoc network in which active sensor-nodes transfer their information to a collection point, or network gateway, by hopping the data through a sequence of nodes. The research has addressed issues such as the optimum routing algorithm to collect the data, and how to make best use of the energy available at each node. These topics are often related because a network route might need to change as the result of a failing battery at an intermediate node. Some researchers have proposed that sensor networks may extend the reach of the Internet by allowing web-based clients to make queries about the physical world. Wireless ad-hoc networks deployed on a large scale have the potential to provide that service by bridging their data to the Internet at a gateway node. However, sensor networks require power if they are to operate for an extended period, or else batteries need to be used and replaced periodically, something that is too costly and labor intensive to consider on a large scale.

Long-range RFID sensing can provide a partial solution. The same principles used by far-field RFID can be used to build sensors that are powered typically up to 20 feet, and communicate data back to the reader using backscatter modulation. At Intel Research Seattle the Wireless Identification and Sensing Platform (WISP) initiative sets out to explore the limits of long-range RFID sensing. A WISP is an augmented RFID tag designed so that it can reside in the interrogation field of a reader and accumulate more energy than is needed to perform a simple ID function. The additional energy can be used to power a microcontroller with an A-to-D converter, allowing a sensor to communicate its state. The energy reaching a tag from a reader will approximate to a $1/(\text{distance})^2$ function, and therefore at 20 feet the energy available is small. The availability of the sensor will be determined by the relative time spent reading the sensor's state in relation to the time it spends in the field between interrogations.

Some of the first motion sensing WISP designs [36] required no more energy than a conventional RFID tag. Simple passive motion-sensors, provided by two miniature mercury switches were used to select between two unique ID-chips, independently connecting each one to a common antenna. The mercury switches can be physically mounted 180-degrees with respect to each other so that as a WISP is inverted it will switch its ID, and as a result, the WISP tag can sense tilting or rotational motion. If it were attached to a vertically mounted spinning wheel, the rate of alternation between the two IDs, as determined by a reader continuously interrogating the tag, would measure the rotational speed of the wheel; as long as the centrifugal acceleration of the wheel is less than the acceleration due to gravity.

Zero-power sensing is limited to a small set of sensor technologies (such as mercury switches), but most of the more diverse sensors require power. The later WISP designs [37,21] use an energy scavenging circuit that can efficiently store enough energy to power a sensor, plus the associated conditioning circuits, and a low-power microcontroller (Fig. 6.5) which can also provide digital filtering, a coded representation of the measured parameter over time, and error-control coding.

Using this approach, long-range RFID sensors (20 feet) can be used to extend ad-hoc sensor-networks another 20 feet into areas where it may only be feasible to provide power to

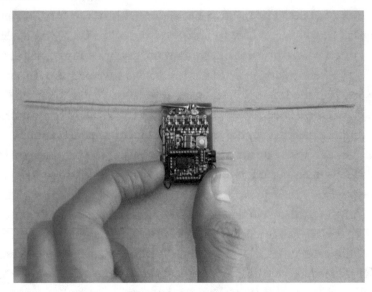

FIGURE 6.5: Showing a prototype power scavenging WISP tag

one central node (the tag/sensor reader). For example, if RFID strain gauges were placed in an engineering structure such as a bridge to determine its response to load, using conventional designs, cables providing power and sensing would be needed to connect to each sensor. Using WISPs, only a few tag reader hubs would be necessary to interrogate the outlying sensors in the extended locality.

Other applications of long-range RFID sensors include: warehousing (measuring conditions inside crates high up on storage shelves), monitoring long runs of pipes, and radiation/chemical detectors inside a transport container.

6.8   SUMMARY

As RFID becomes more prevalent, growing economies of scale will allow environmental sensors to be integrated with tags allowing RFID readers to report back on a wide variety of real-world conditions. Tag readers that are typically powered by the mains, or conventional batteries, will in many cases be able to access wireless networks connected to the global Internet. They will be able to relay the state of the physical world, and make it available to users and servers through web-services, allowing real-time data mining to be implemented on a scale larger than ever before. RFID sensing is part of the classic story of Ubiquitous Computing [27], a technology that can improve our lives while remaining invisible, as users do not have to be directly aware of it to reap the benefits.

CHAPTER 7

Deployment and Experience with RFID Systems

This section describes trials of RFID systems that are breaking new ground by utilizing RFID to support commercial ventures that have not previously adopted this technology. We summarize three high-profile systems that provide insights into the benefits of modern RFID applications:

- "Store of the Future"—Metro AG;
- Wal-Mart's trials of RFID;
- Frankfurt airport's maintenance operations.

Looking further into the future, we describe two additional deployments from the realm of research that provide a glimpse of the variety and scope of future tagging systems:

- iBracelet: supporting work practices;
- University of Washington plans for Ubiquitous RFID reader deployment.

7.1   STORE OF THE FUTURE—METRO AG

One of the best known, and most progressive, trial sites for testing RFID concepts in the retail trade is the "Store of the future" established in April 2003 by the Metro AG group in Rheinburg, Germany (the world's fifth largest retailer) [38] (Fig. 7.1). Financial support has also been provided by major IT companies such as Intel, Philips, SAP, and IBM who strive to learn how RFID might change processes in the retail trade, and how to integrate this technology with existing supply chain management systems, and the computer systems that support them. Metro's main objective is to improve the management of its goods at all stages of the supply chain, from a supplier's warehouse to the shelves of retail stores, utilizing the efficiency of automation wherever possible. In addition, the store provides an opportunity to explore applications beyond the supply chain, providing a richer and more engaging experience for the customers, enabled by the use of modern RFID tags and other state-of-the-art technologies.

FIGURE 7.1: Metro—a future store used to test new RFID concepts

Customers experience the technology as soon as they enter the store pushing a *smart cart* that has a tag reader built in. Soon they pass *smart shelves* that also contain RFID readers, and *smart checkout* aisles that speed the process of payment. The smart carts also contain a tag and their identity recorded as they enter and leave the store. The number of carts in the store can be used to gauge how many checkouts need to be open for optimal efficiency. As the customers remove items from smart shelves they automatically inform the store management system about the goods that need to be replaced and where they need to go. A customer can also scan items as they are placed in their cart to keep a running tally of the purchase cost (initially only some items were tagged with RFID and barcode scanners were also used). At the checkout, the cart's RFID tag is further identified and the grand total displayed for payment.

Specific suppliers such as Gillette and Proctor & Gamble were early collaborators, but commercial interest has been growing, and by 2006 over 300 suppliers for the Metro Group are expected to incorporate RFID in their delivery pallets as part of this trial. There continues to be some debate over the benefit of pallet versus item tagging. The greatest benefits initially come from pallet tagging, but item tagging is likely to follow once the cost of the tags drop further. RFID labels on individual items can also serve as an antitheft mechanism when used in combination with readers at store exits. By comparison, conventional UPC codes do not provide any form of antitheft detection, and if products need to be protected in this way, they require an additional antitheft tag. RFID can serve as an identity, an antitheft tag, and even provide a tamper detection mechanism (see Section 6).

At the Metro Store of the Future, RFID has also been used to test new services that help customers choose merchandise they may wish to buy. For example, tagged CDs can be passed over a reader station attached to a computer, and a snippet of the music played back without opening the CD package. Food that may have an expiry date can be interrogated as it is purchased to ensure it is still fresh enough for consumption. And tagged clothes can be scanned in front of a monitor, showing the customer how they look on a model. Several items scanned in this way allow combinations of clothes to be seen together in order to quickly understand how they blend, and all without trying anything on. These systems can also make recommendations for alternative garments that might replace, or be added to, existing choices. Even in a dressing room this system can have added value by alerting staff to bring new garments, or additional sizes, to help a customer find the items they are looking for. The jury is still out on the value of these services, and many others in the planning stage, but the Metro trial does allow these ideas to be tested and move beyond idle speculation.

Reports from Metro in early 2006 indicate that efficiency and cost saving are already apparent, and 250 of its 2300 stores in Europe and Asia are in the process of installing RFID-based supply chain management systems.

7.2   WAL-MART RFID TRIALS

Wal-Mart, the largest retailer in the world, has been one of the driving forces spurring the RFID industry to provide effective solutions to improve the efficiency of supply chain management (Fig. 7.2). Initially, the deployment of RFID technologies was limited to tagging pallets and cases, and not individual items, as it was believed that the cost savings in this area alone would provide enough justification for the project. In 2004, at an early stage of deployment of EPC-Global tags in their supply chain, the measured benefits were described as "promising."

FIGURE 7.2: One of Wal-Mart's early RFID trials stores

By October 2005, the results reported were more quantifiable, with data captured during a 29-week trial period in 2005, using 500 of its Wal-Mart and Sam's Club stores, supported by 140 suppliers. The results can be summarized as:

- 16% decrease in out-of-stock items for EPC tagged products; and
- 300% improvement in restocking time for EPC tagged items in the store.

It is further reported that at the trial sites excess inventory was also lower, although not quantified at this time. Riding on this success, by Q1 2006 the number of Wal-Mart suppliers incorporating RFID is expected to increase to 300 [39]. The RFID industry has been fortunate that Wal-Mart has chosen to lead the charge. If it were not for its premier market position, a lesser company would probably not have the economic power to persuade suppliers to adopt RFID, which at first is only going to benefit the retailer. Supporting Wal-Mart's position, Tesco, the UK's largest retailer (the world's third largest) is also pioneering the adoption of RFID. In the US, the DoD is another major consumer of goods and materials with its own unique set of supply chain requirements, and it too is carrying out trials based on EPC RFID tags. In summary, there is considerable momentum behind RFID standardization and adoption by the major retailers of the world, and is likely to have a knock-on effect for smaller companies.

Having the first quantifiable numbers that demonstrate real value in the supply chain will also provide a foundation for the RFID-based services being investigated in the Metro AG stores. From a customer perspective these are likely to be more exciting than improving the supply chain, but would not have been a large enough economic driver to motivate RFID adoption on their own.

7.3   RFID SUPPORT FOR MAINTENANCE OPERATIONS AT FRANKFURT AIRPORT

The second largest airport in Europe is Frankfurt, typically handling over 50 million passengers per year, and in some ways can be compared to a small town, relying on numerous utilities and support services for daily operation. Although there are many well known applications of RFID in an airport, such as tracking airline-baggage as described in Section 5, there are many other lesser known procedures that can benefit from it. For example, electrical systems require regular maintenance to provide uninterrupted operation and ensure the safety of passengers. The University of St. Gallen have studied how RFID has been successfully trialed at Frankfurt airport [40] to improve the efficiency of maintenance operations, and their observations are summarized here:

The airport maintenance operation group, Fraport, is responsible for ensuring smoke and fire control systems are always fully operational. The danger of not doing so is highlighted in the

1996 Dusseldorf airport fire that resulted in loss of life. Following this tragic event, to improve on safety the German government created new legislation that mandated the use of improved and more extensive maintenance logging to ensure essential work would be carried out in the future. Given the cost constraints of maintenance organizations, some of their activities are carried out by external firms and it is necessary to set up procedures that can efficiently validate and record the maintenance process, which is made more difficult as the work force is rapidly changing. The airport authorities have a responsibility to ensure this work is performed according to government regulations, and in the event of an actual fire the authorities would be negligent if proof of maintenance was not available.

In 2003, Fraport, decided to use RFID (Fig. 7.3) to support the inspection and maintenance processes required for the fire shutters and ventilation system installed throughout the airport. The system made use of RFID by deploying a tag next to each of the fire shutters. A maintenance technician, supplied with a handheld computer incorporating an integrated RFID reader, checks each shutter and uses the computer to log each associated tag, thus providing evidence each fire shutter location had been visited during the inspection process. For each scan of a tag the computer engages the technician in an electronic dialog to ensure that every aspect of the inspection/maintenance has been considered and the answer recorded. At the end of the dialog, the technician is asked to scan the tag once more, this time writing a reference to the computer log into the tag's memory, along with the date and time of the inspection. This process locks the log file so that it cannot be changed at a later time, and provides two independent points of reference for the inspection, the RFID tag and the computer.

Since the electronic log replaced a previously handwritten report that was often incomplete, and resulted in 88,000 pages of logs per year (for 22,000 shutters) which also needed to be archived for 10 years, the system was a considerable improvement and provided tangible cost saving. Furthermore, it is reported that the technicians are now more motivated, preferring this process over handwritten documentation. Based on this positive experience, Fraport is likely to extend RFID supported maintenance to other aspects of the fire prevention system, and to tracking mobile equipment that is frequently misplaced within the airport campus.

7.4   INTEL RESEARCH: IBRACELET AND DETECTING THE USE OF OBJECTS

Intel's Research Laboratory in Seattle (IRS) has been experimenting with systems that improve work practices, and enable the elderly to live independent lives in their own homes for longer than would have been previously possible [41]. The thesis behind this work is that if you can understand what somebody is doing through their actions, you can automatically provide help when a problem arises. In the case of the elderly, it may also be possible to use a log of their actions to determine if a person's mental ability is stable or in decline.

FIGURE 7.3: RFID maintenance tag—also includes a barcode for redundancy

In order to support this exploratory work, a system must be built that can accurately and automatically capture events in our daily routine both at home and at work. Although this could be a task for computer vision, placing RFID tags on all the objects that need to be monitored, and employing wearable tag-readers to log the objects used by our test subjects, is an alternate approach to the problem. The tagging solution also has some advantages in that the processing requirement for the system is reduced, and the accuracy of interpretation is much greater. Furthermore, cameras raise many privacy concerns, whereas reading tags in the environment is less intrusive with arguably better fidelity.

Since we interact with most of the things in our lives by touching them with our hands, an RFID reader was designed to be worn on the hand, and record nearby tags. During the evolution of this concept the initial implementation consisted of a modified glove, the iGlove, which incorporated a short-range reader on the top side of the garment. However, although successful as a reader, the glove was uncomfortable to wear indoors and impractical in many domestic situations and so an improved follow-on design was created based on a plastic molded bracelet—the iBracelet [42, 43] (Fig. 7.4). This device was more convenient to wear, but mounted further away from the fingers compared to the iGlove solution. As a result a major consideration of the design was how to increase the range of the reader from the previous 2.5 cm to 30 cm (the reference says 10 cm but the RF design has improved further since then), without significantly increasing the power consumption of the device, and hence without decreasing its battery life. This problem was overcome by improving the design characteristics of the antenna and associated load demodulation.

FIGURE 7.4: The iBracelet created by Intel Research Seattle

The iBracelet is currently being considered for trials within Intel's silicon fabrication plants in order to track the progress and management of silicon wafer carriers at the plant—a valuable commodity that when misplaced, or more seriously process steps are left out, can result in a considerable loss of revenue. Other projects at fabrication plants, such as the LotTrack project at Infineon, have already found value in tracking wafer cassettes using RFID [44]. iBracelet is also expected to provide support for community "Aging in Place" projects, but at this stage it is being used to verify that statistical models can be constructed that produce accurate recommendations in a test environment before being trialed in the real world.

7.5   UNIVERSITY OF WASHINGTON'S RFID ECOSYSTEM PROJECT

The University of Washington's Department of Computer Science & Engineering is in the progress of deploying a large experimental RFID system in their new building, the Paul G. Allen Center on the UW campus (Fig. 7.5). By deploying ubiquitous RFID readers mounted at doorways, ends of hallways, and other pedestrian funnel points in the building, it should be possible to monitor the comings and goings of many different types of tagged objects. The objective of the project is to investigate a host of consumer (as opposed to supply chain) applications of RFID such as reminding, finding lost objects [45], inventory control, gaming, activity inference, etc. At the same time, the focus is on building a data architecture that will respect user privacy and allow users to retain control over how their RFID tags are utilized. To that end, the research is investigating approaches for encryption, selectively enabling tags for specific applications, and using readers that act as local RFID sensors and rebroadcast their tag

FIGURE 7.5: Paul Allen CS Building at University of Washington

reads rather than sending them to a centralized database. The project is currently funded by NSF and the UW's College of Engineering with support from Impinj, Inc., a leading manufacturer of EPCglobal Generation-2 tags and readers that enjoy the benefits of long read-ranges, fast multitag singulation, and on-tag memory.

7.6 FUTURE DEPLOYMENTS

Looking beyond the current commercial applications, and the new trials in progress, RFID technologies and large-scale business applications are still in their infancy. There are some problems to overcome, but the technology is versatile and can be adapted along many dimensions to provide effective solutions. If this text were to be revised in the future, the scope of this section with respect to deployment and learning, will be likely evolve considerably.

CHAPTER 8

Privacy, Kill Switches, and Blocker Tags

One of the reasons RFID has been written about so much in recent years is that some people believe the introduction of RFID technology will erode their right to privacy [46]. Privacy advocate groups are concerned that even though many of the corporations considering using RFID as part of their inventory tracking mechanism have honorable intentions, without due care the technology might be unwittingly used to create undesirable outcomes for many customers. The inherent problem is that radio-based technologies interact through invisible communication channels and we are not aware when communication is taking place. Consider a situation in which RFID tags are used to label garments in a clothing store. From the store's perspective a conventional inventory stock check is difficult because customers frequently mix-up the garments, and theft can take place making the sales records incomplete. On the upside, with RFID tagging the various racks and bins of clothes can be checked very quickly, even when muddled, improving the efficiency of the store. On the downside, if a tag is not removed when a customer buys an item of clothing and later wears it, the tag can be used to track them wherever they go. This capability might be used by other vendors to learn about the shops they frequent, and then target them with direct marketing based on this information. This scenario was presented graphically in the science fiction movie 'Minority Report' in which the hero, played by Tom Cruise, was identified in department stores not by RFIDs in his clothes, but by his eyes and the use of ubiquitous retina scanners. As a result he was subjected to a multitude of multimedia marketing materials chosen to appeal to his lifestyle (see Fig. 8.1). In this story in which he was trying to avoid arrest, the solution was more dramatic than removing tags from his clothes, he had to find a surgeon that could perform an ocular transplant. While removing RFID tags from clothing purchases would not be so dramatic, it would be very frustrating to have to take this kind of action in order to maintain personal privacy.

In an even scarier scenario, criminal elements could judge your personal wealth depending on the purchases you have made, and then target you for theft. It was because of a growing cloud of public and media concern that Benetton, a well-known clothing store, had to make a hasty

FIGURE 8.1: Scene from the movie "Minority Report" in which billboards customize themselves to the shoppers in the vicinity

retreat after it announced its plans for using RFID tags in its stores [47]. A similar response resulted from the US government's plans to put RFID tags into passports in an attempt to make them easier to check at borders, and harder to forge. However, privacy advocates would argue that covert readers might steal information that can be used to enable identity theft [48]. In contrast to Benetton, the passport scheme is still going forward, although its implementation is being modified to address some of the public concerns.

This potential for misuse of RFID is high, but like many modern technology debates the story often has two sides. The undesirable scenario relating to tagged garments described above can be turned into a potentially useful one. Washing machine manufacturers could integrate RFID readers into the door of their machines, making the machines aware of all items that have been selected for washing. As a result, they could choose the appropriate washing cycle, and possibly warn you about incompatible garments that might result in color runs.

8.1   KILL SWITCHES

In order to overcome many of the concerns, EPCglobal designed a feature into its RFID tags called a *Kill Switch*. This allows vendors to permanently disable an RFID tag at the point it is sold, without necessarily having to remove the tag itself, which might be woven into a garment deliberately making the tag difficult to remove, and serving as an antitheft device. Kill switches can certainly help, but there are additional concerns that retailers may become complacent, and that not all stores will be vigilant about disabling the tags. It is still possible that an insidious number of operational tags could hitch a ride in our clothing and later on, criminal elements could take advantage of this situation.

8.2   BLOCKER TAGS

RSA corporation has proposed a solution that individuals can take into their own hands, the concept of a *Blocker Tag* [49]. This is a modified RFID tag that takes advantage of the anti-collision protocol used by EPCglobal Generation-1 tags by responding to each interrogation in such a way that it appears that all possible tags are present. As a result the tag reader has no idea what tags are actually near by. Perhaps having simple countermeasures to prevent misuse of these tags is exactly what is needed to overcome privacy concerns.

8.3   TAGGING IS ALREADY AN INTEGRAL PART OF MODERN LIVING

Taking a more general view of electronic tagging; cell phones, credit-cards, and networked computers, similar to RFID, are all technologies that make use of a unique identity which is a fundamental part of their operation. Because of its uniqueness, the number can be used to identify and locate each instance of the technology, and furthermore, because these items are also related to chargeable services, the service providers are able to track the locations and activities of their customers. So, how are RFID tags different and should we be any more concerned about them than computers, cell phones, and credit cards that are already part of our lives?

One difference is that all these devices provide considerable utility in a modern world, and, so far, we have been prepared to give up some amount of privacy to enjoy their benefits. As with most technologies there are advantages and disadvantages, and we must individually evaluate whether the benefits outweigh the cost. On the other hand, RFID tags embedded in the merchandize we buy have no direct value to us. In fact, on the downside, undesirable applications such as personalized marketing campaigns are the most likely result. Perhaps it is the imbalance of the technology's pros and cons for the consumer, in favor of the cons, that has contributed to antagonism toward RFID tagging.

When using cell phones, credit cards, and computers, we inadvertently give up information about who we speak to, our location, what we like to buy, where we buy it, and when we surf the world-wide-web we also give away our personal reading preferences. We continue to do this because mostly this information is kept confidential, and no detrimental consequences result. We should ask ourselves if this will be true for RFID. For some technology we can take control of the privacy issues by carefully choosing how we use it. For example, credit cards give away our location each time we make a purchase, but as the cards are personal and we control the account, we can explicitly decide when and where to apply them. The use of cash is always an option, and just knowing there is an alternative makes credit cards more palatable.

Cell phones are more problematic in this regard because they do a poor job of protecting location privacy even when we are not making a phone call. Simply put, a wireless service provider needs to know your location at least to the closest cell tower, in order to route your phone calls. When carrying a cell phone that is turned on, we are continuously giving away our location without explicitly intending to do so. We can choose to turn the phone off, and only turn it on to make a call, but then we lose all of the advantages of being able to receive calls from friends and colleagues. Further, the short-range radio technology Bluetooth, which is also integrated with many cell phone products, contains a unique MAC address to support its protocol. And when turned on, it is possible to track your location using nearby computers that can automatically discover the device. The Bluetooth radio can be turned off manually, but many people are not familiar with this cell phone capability and simply may not realize it is a privacy concern.

Identity theft is also a danger when our identity can be defined by a single unique number. Even if we are personally unconcerned about technologies that reveal our identity and location, perhaps because we trust the service providers and feel we are doing nothing wrong, there is the potential for nefarious individuals to steal our identity. For example, if our cell phone or credit cards are stolen, the thief will appear to take on our identity and journey to places and initiate transactions that are beyond our control. In the case of a phone or credit card, which is used on a daily basis, it is likely we will discover the loss early on and report it to the service provider or bank, before too many charges accumulate. However, we might still need to dispute a bad credit report or a large phone bill, if the theft is not reported promptly, and therefore must take special care to keep track of these items.

In the case of RFID tags, identities may be stolen inadvertently, not because anybody wanted to steal a tag, but because the tag was embedded in a high-value item that had been associated with a person at the time of purchase. In a world in which tags and readers become more ubiquitous, a thief carrying your purchased items may inadvertently link you to a crime scene. In one vision of the future, if such inferences can be easily made, it may be necessary to keep track of all personal items that contain RFID, and report them when stolen. This has an upside and a downside. The upside is that stolen items may be more easily recovered when you report the theft and the thief will be caught. Or the downside, you do not report an item stolen, and your identity is falsely associated nefarious events. However, although inference based on RFID, or other forms of electronic identity, may cast suspicion, without additional proof it is likely we will always be able to claim "plausible deniability," and further proof will be required. It should be noted this concern is not without precedent, as in the past car license plates have been used to infer the identity of people attending illegal events, simply by noting the license numbers of cars parked nearby. However, innocent people who happen to park nearby also become suspects. False inference of this kind provides us with reasonable cause for concern about the potential secondary uses of RFID tags.

8.4   FUTURE IMPACT ON SOCIETY

In this section we have considered the pros and cons of using RFID and why concerns arise. Looking forward, how will this technology be deployed and will the concerns be taken into account to create the appropriate legislation? Is it also possible that our current expectation of person privacy will soon to be lost?

RFID is just one of many new technologies that are able to track our location and automatically identify us. Consider how many traffic-cams installed along roads can record your car's license plate as you drive to work. Given the application of technology in all aspects of life is continuing to increase, it is worth considering how society can influence the use of emerging tagging technologies, and how it might adapt to their adoption.

Opposition to RFID tagging is most likely to be strong in the labor force that supports industrial manufacturing. Due to competition that arises from globalization, this industry is forced to remain competitive at every level, a problem undermined by the labor costs in the industrialized countries. Tagging provides a tool to monitor work practices and improve efficiency. If deployed in moderation, the results are likely to positive but if too aggressive, may become onerous for workers. The latter is clearly a future we should try to avoid. On the other hand, for example, if one person is able to rapidly record the inventory of an entire warehouse, using a motorized cart, a computer and an RFID reader, something that previously would have only been possible with a team of workers, this is a level of progress that industry cannot afford to ignore. Society, on the other hand, must then come to terms with more job losses, and the retraining of workers to create skills that are in demand.

As a further illustration of the reaction to electronic tagging, in 2002 when Tesco in the UK started RFID trials at one of its grocery retails stores in Cambridge, the result was open protest outside the store. Figure 8.2 shows one of the protesters holding a poster with the slogan "Say 'No' to Spy chips." The concerns are real and should be taken seriously if the technology is going to be deployed effectively without a consumer or worker backlash.

In the US, currently, there are pressures on the government to gather internal intelligence in order to combat terrorism. After the tragic attacks on the World Trade Center in New York, and the Pentagon in Washington on 9/11 2001, there is well-founded concern that terror groups waiting for similar opportunities are already within the country. The ability to keep track of electronic identities could play a role in helping the homeland security office do their job. By gathering information about events involving mobile devices, people, and places, and continuously feeding this data into computational inference engines, it may be possible to look for suspicious circumstances that need further investigation. RFID may also play a role in these investigations. Because repeating the circumstances of 911 is something that we all wish to avoid, it could be argued that the government has the right to track electronic identities in this

FIGURE 8.2: (a) Tesco trial store in Cambridge, UK. (b) Protester outside Tesco in the UK voicing concerns about RFID trials

way, and to some degree should be a counter-balance to the concerns described earlier. Just as airport security checks are an invasion of privacy and often inconvenient, it is something we are all prepared to endure, because the consequences of poor airport security are too terrible to leave to chance.

In a future world in which RFID has been deployed on a grand scale, it is possible there will be databases that record our identity along with the time and places we go from day to day. This information can be protected and used for the benefit of society, or be made available indiscriminately, and it will be up to government policy to protect us as these policies evolve. In Europe, "The Data Protection Act" already limits access to all computer records that contain private information requiring written consent for its disclosure, but similar legislation does not exist in the US at present. As RFID technologies mature, and the EPCglobal standard becomes adopted around the world, the issues surrounding the use of electronic tagging will become better known, and it is likely there will be more international agreement on when and where this type of information can be disclosed.

CHAPTER 9

Opportunities for RFID Integrated with Memory

A distinguishing feature of modern RFID is that electronic tags can now contain far more information than a simple identity. Today, it is possible to integrate additional read-only or read/write memory into a tag, which can be queried or updated by a tag reader/writer. If, in the future, RFID becomes an established technology that is used to label common products, distributed read/write memory in these devices could become a resource that application developers will use to their advantage [50]. Here we discuss some of the possibilities for this distributed memory revolution.

9.1 READ-ONLY MEMORY

When using RFID tags to identify consumer products, additional read-only memory in the tags can store product details. This information does not need to be read every time a tag is interrogated, but is available if required. For example, the tag memory might contain a batch code, and if some products are found to be faulty, the batch code can be used to find other items that potentially have the same defects.

An alternative approach to tag-based memory is to use a tag's unique ID as a key into an online database to recover the batch code, and other product specific data. However, there are many situations in which communication with a database may not be possible. For example, the store selling a product may not have access rights to the computer systems used by the manufacturer. By writing the batch number directly into the tag, it is accessible at all stages of the supply chain. Consider another example involving a tagged parcel that is misdirected during transportation; the receiving organization may not be able to determine its intended destination. Additional information in the tag can be self-describing, and include the name of the destination stored as a human readable text string within its memory, thus obviating the need and cost of a fully networked tracking system.

Although today's passive writable RFID tags can only store up to approximately 8000 bits, in the future, tags may have much larger memories with megabits of data. Taking advantage of this opportunity, manuals, or other documentation associated with a product could be stored in the same tag that also identifies it for sale. Further, by embedding the tag into the product's case, it cannot be easily separated; with the result that the documentation will remain readily available. This is an advantage because paper manuals are frequently lost, leaving the owner in some difficulty when trying to find out how to use an unfamiliar product feature.

Similar to the batch-code example given above, finding a product's manual also has a network-based solution. The manufacturer can provide a consumer with an online manual through the world wide web. However, this approach may not stand the test of time as the consumer must now rely on the manufacturer to maintain the website. In practice, modern products have short lifecycles, and the companies involved can fail financially, thus the web-based information might disappear even though the products are still in use. RFID-based memory embedded in the product does not have this shortcoming. Furthermore, it can also help conserve natural resources by lessening the environmental impact associated with creating extensive paper documentation.

9.1.1   Enhancing Objects with RFID Memory

There are many examples of how RFID can usefully augment interactions with our environment. Consider a poster advertising a movie at a nearby theater (see Fig. 9.1). It will likely contain a title, a graphic that depicts the story and the characters, a list of actors, and below that a date

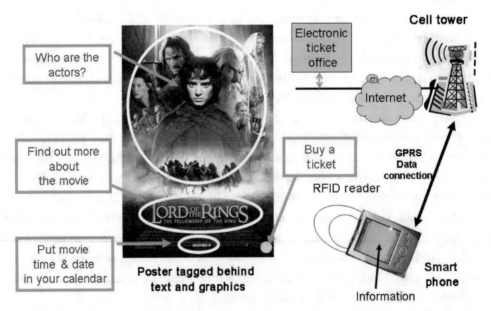

FIGURE 9.1: A movie poster with additional information provided by embedded RFID tags (poster © Newline Cinema)

indicating the opening night. RFID tags can be attached to the poster behind each of these regions in order to provide additional information to a curious passer by. For example, the poster may have intrigued you, but you may want to learn more about the movie before deciding to see it. A pocket computer, or smart phone, with an RFID reader can be used to interrogate the tagged title region and obtain a detailed summary of the story. Moving the reader over the pictures of the actors would provide information about other movies they have starred in. And moving the smart phone over the text of the movie's opening date and time, could instruct your smart phone to open your electronic calendar and enter the movie title into your schedule at the corresponding date and time. This example illustrates two points that have not been discussed earlier. First, an object may have multiple RFID tags attached to it, the purpose of each being indicated by its position on the host object. Second, additional instructions stored with the data can suggest how to process it. In the last example it was processed as calendar information. This technique can also be used to initiate financial transactions. Extending the poster example, there might be an additional area that advertises "Buy Ticket Here." Scanning this area with a smart phone supporting both an RFID reader and GPRS capability, could automatically and wirelessly connect the phone to a ticket office provided by an internet service, and purchase the ticket electronically. To mitigate accidental purchases, when the phone is unwittingly brought too close to the poster, the system should provide a confirmation dialog before committing to the payment.

9.2   READ/WRITE MEMORY

More intriguing applications of RFID take advantage of read/write memory available in some types of tags. Writable memory sizes are likely to follow similar trends to that of read-only memories. However, as data can be stored in arbitrary formats by an interrogator, its use is only limited by the creativity of the application developers.

For example, secondhand consumer goods which contain write-once embedded RFID tags may tell you something about the prior list of owners and when and where ownership changed hands. This is similar to the provenance documentation that usually accompanies valuable antiques. RFID tagging may extend this kind of tracking to everyday items, allowing consumers to have greater confidence that they are making good purchases and the price of the item will be reflected by its history. As a result there is the potential to have a higher resell value when the provenance is favorable, which may motivate buyers and sellers to ensure these records are correctly maintained.

One of the consequences of using RFID for most forms of automatic identification is that in time it may lead to an extensive deployment of electronic tags in our surroundings. If these tags contain digital read/write memory, our homes, cars, offices, and cities could soon have the memory resources to store sizable amounts of data. For example, if public places are tagged with RFID [51], this can be a resource available to the city planners to store location-based information accessible by the public. Such memory might be used for recording historical data, or information about community services available in the locality, or provide data about the presence of utilities, such as electrical cables, gas, and water pipes. An advantage of RFID read/write memory is that its use can be decided after its deployment, and thus city planners only need to ensure memory-based RFID is put in place during construction or renovation, and decide on its content and application at a later time. Furthermore, information can be written to these tags using a simple handheld interrogator, and thus maintaining the information is no less burdensome than any other city maintenance task, such as checking conventional signs are up-to-date. RFID-based information in this context can also save the city money when compared with the cost of creating physical signs that must be large and made of materials robust enough to endure severe weather. Furthermore, updating physical signs is a costly endeavor, but updating information stored in an RFID memory is inexpensive.

RFID memories can also be made available as a medium for public messaging. In the same way that message boards are available at the entrance to some campuses, you might be able to leave messages for friends stored in RFID tags mounted on walls for the same purpose. These messages could be accessed by a handheld-reader based on a smart-phone using the NFC standard. One can imagine a creative younger generation having fun with this concept: It could even lead to a new type of graffiti with a more socially acceptable outcome than the property damage resulting from the conventional form.

9.2.1   Location and Directions

In a city that in the future may deploy RFID tags on signs, street corners, and building placards; tag IDs can be used to help determine a location or the direction of travel, and are an inexpensive alternative to a portable Global Positioning System (GPS). Smart phones which already support enough memory to store digital maps can augment this information with databases of tag IDs tabulated against location coordinates and lists of nearby businesses. This data can be used to graphically illustrate your location on a map, and help you locate nearby shops and restaurants. For example, you could scan an RFID tag on a street corner and then enter a query, "Find the nearest Chinese restaurant." By keeping track of a sequence of tags along the way, the mobile computer could also determine the direction you are walking, and if incorrect, provide updated instructions about the route.

9.2.2   Memory and Time

RFID tags with embedded writable memory can be used to log information, but unlike a computer file system, they do not have the ability to automatically timestamp the data unless the interrogator itself can provide timing information. This is because a clock requires continuous power to keep it operational and once an RFID tag leaves the reader's field, energy is no longer available. However, a timestamp is important for many applications, allowing data to be merged and synchronized, and it can be used to prevent data-logs from being falsified. An example application of writable RFID tags along with interrogator time-stamping, is provided by the sport of Orienteering, and is described below:

Orienteering

The sport of orienteering, although not well known in the US, has a strong following in Europe, and is beginning to grow in North America. The sport combines cross-country running with map and compass-work. At the start of a race, contestants are presented with a detailed map marked with a route containing way-points (also called controls), the locations of which are only previously known to the organizers. A runner must visit each control in the order shown on the map and complete the course in the fastest time in order to win the race. The starts are staggered to make it less likely runners will follow each other. Traditional orienteering races supply runners with a course card and use mechanical punches at each of the controls; a runner must punch the card at each control to prove that he or she has been there. Each punch contains a unique pattern of pins, which is also not known to the runner, and therefore effectively unforgeable before the race. A modern orienteering event provides runners with an alternative to the card, a finger mounted RFID tag (see Fig. 9.2a) that can be inserted into a tag writer housed in a small battery-powered box at each control (see Fig. 9.2b). As a runner visits each control, the tag accumulates a set of unique numbers identifying the controls, and the time they were recorded. This information can be downloaded at the finish to satisfy the race officials that the course has been completed correctly, and to determine the total running time. As with the marathon races described in Section 5, the "split times" calculated for intermediate points along the course are

FIGURE 9.2: (a) RFID thumb tags. (b) A control in the forest with RFID writer station and position to insert the thumb tag

also of great interest to the contestants. At the end of the event, the tags' memories are wiped clean, ready for the next race.

9.2.3   Another Use of Time

Even though read/write RFID tags must rely on an interrogator to generate timestamps, it is still possible to use the parameter of time to police good behavior. For example, interrogator timestamps can be used to help detect falsified events. Consider two interrogators that create write events sequentially, each providing their own timestamp. The time recorded for the first event constrains the earliest time that can be recorded for the second event. Thus, if the second interrogator wishes to falsify its log, events that occur both before and after its own recording provide limits on the time of the forgery.

The value of this technique can be illustrated using a supply chain example. Consider a tagged case of commercial goods that are being transported by a shipping company between two cities, and along the route the goods must pass through several checkpoints. However, at one checkpoint a nefarious operator decides to remove some of the merchandise, and to cover his tracks tries to falsify the time that the goods were in his hands. An automatically generated timestamp in the tag would prevent this type of forgery. But, as this is not possible with passive RFID and because the timestamp must be provided by the interrogator, the nefarious operator has the opportunity to change his timestamp at will. However, if each event is recorded into a write-once memory in the tag, the nefarious operator can only claim his event occurs after the previous write event, and cannot project the time too far into the future, or it might conflict with the checkpoint that follows. RFID tags that create a progressive log of data using an append-only memory can therefore be used to detect some types of supply chain anomaly without the need for an active clock in the tag itself.

9.2.4   Facilitating Wireless Connections

In Section 3.4 we described how NFC could be used as a side channel to aid wireless discovery as well as serve as a communication channel in its own right. Bluetooth and WiFi are standards for localized wireless communication (WLAN) that can benefit from this capability. RFID memory without any special extensions can also serve as a side channel to aid in communication by passing auxiliary information. Mobile devices can use this information to decrease association time and reduce power consumption. For example, if mobile device A would like to connect to another device B that uses a short-range RFID tag to store its wireless MAC address, it can discover and connect to it by reading these parameters when brought close by. If several other devices are present in the locality (C, D, and E), A no longer needs to contend for the wireless medium in order to discover them all and then decide which one to connect to. A wireless discovery process is inherently unreliable because a mobile device does not know what other devices are actually in the locality, and therefore the value of retrying a discovery request that had a null result is of limited value. That is to say, the messages may have been lost due to wireless interference or contention, or there may not have been any other devices present at all. In practice engineers design a wireless discovery mechanism so that a statistical argument can be made about the likelihood of a successful discovery, e.g., for the Bluetooth protocol [24] the typical discovery process is carried out for 10.25 s with a greater than 95% chance that all nearby devices will be found. For Bluetooth, an RFID side-channel can improve on this long discovery time, by allowing users to deliberately bring devices together in close proximity in order to initiate the wireless link. By involving users and making use of unambiguous physical-proximity, wireless connection time can be shortened and the results accurately reflect the user intention.

Power savings are also possible using an RFID side channel. Radio standards such as WiFi were not designed with a low power mode that enables them to be discovered without the radio being turned on. This results in a considerable quiescent power consumption, which for small mobile devices dramatically shortens battery life. A modified RFID tag can serve a wake-on-wireless capability [52] for WiFi or other similar radios. This is achieved by building an electronic switch to turn the WiFi radio on and off, and controlling the switch from a modified RFID tag. In short, the RFID circuit can be extended to provide an external (wake-up) signal, the logical state of which is defined by the value of an internal memory register written by an RFID interrogator. Thus, an RFID interrogator can wake-up a near-by WLAN radio by writing into the tag's memory.

In addition to using RFID to pass MAC addresses and wake-up information, there are higher levels of interaction that can be usefully communicated through its memory. The discovery mechanisms described above are specifically related to wireless hardware. In the case of discovering a simple device, such as a wireless printer, it is expected to provide a print service and nothing else. However, when discovering a general purpose device such as "computer," the list of services available will not be known. In wired networks a service discovery protocol is used to determine the services provided by a networked computer, examples include UPnP and Jini, but these mechanisms assume a shared network connection already exists. In the wireless world a connection must be made before service discovery can be determined. To explain the problem, consider a Bluetooth enabled PDA (A) that is trying to connect to a wireless music service running on one of three nearby Bluetooth enabled PDAs (B, C, and D). The three devices are all capable of advertising their music service using UPnP. By initiating the standard Bluetooth discovery protocol, A will discover B, C, and D, and it will learn they are PDAs and will find the set of profiles available (note: the Bluetooth spec uses the term profile to describe a protocol it supports). Let's assume they all support the PAN profile that allows an IP based protocol to be established between them. In order for A to determine if a music service is available on B, it must create a Bluetooth connection to B, establish an IP connection, and then listen for a UPnP broadcast announcing the music service. If nothing is found, the link must be disconnected and a new connection made to C and D in turn, until the service has been found. This process is time consuming, and power inefficient. RFID memory can save time and power by storing service names and TCP port numbers alongside device types and MAC addresses. If PDA (A) now reads each of the RFID tags attached to PDA (B, C, and D), in a short period of time it can decide which device to connect to, and only then incur the overhead of a connection to the device that can actually provide the music service [53].

9.3   SUMMARY

Through the examples described here it will become apparent that there are numerous applications for RFID memory, many more than can be explored in this article. However, this section provides representative examples to illustrate the general scope of memory applications. Further reading can be found in the cited references.

CHAPTER 10

Challenges, Future Technology, and Conclusion

There are three main issues that are holding back the widespread adoption of RFID: design, cost and public acceptance. So much commercial interest has been building around this technology that adoption is reaching a tipping point and the remaining technical problems are the focus of much attention. New developments are described in the press on a weekly basis, and progress towards workable solutions is likely to be swift. Below we consider the challenges in more detail.

10.1 CORE CHALLENGES

10.1.1 Design

Designing tags and readers so that they guarantee highly reliable identification is not a solved problem. The solutions must be tolerant of tag orientation, packaging materials, and checkout configurations that can be found in typical stores. Improved tag antenna design can solve some of these issues. Tag readers can also be designed to exhibit antenna diversity by multiplexing their signals between a number of antenna modules mounted orthogonally, or by coordinating multiple readers. In the latter case, care needs to be taken to avoid what is sometimes called *The Reader Collision Problem* [54], as interrogation signals will interfere with each other. Multiple readers can be used to provide interrogation diversity if a strict time division scheme is used.

10.1.2 Cost

Pricing plays a critical role in any business decision: Traditional labeling solutions are still considerably lower cost than any electronic tagging solution. Even though RFID tags are now available at prices as low as 13 cents each, the market analysts cannot agree what the tipping-point price might be, and argue among themselves that a 10-cent, 5-cent, or even 1-cent tag is going to be needed before the market begins to cascade into adoption. Consider a 50-cent candy bar—if a 10-cent RFID tag replaces a 0-cent barcode (it can be printed on the wrapper itself), then there may not be any remaining profit. As a result RFID tags are likely to have their first deployments with high-profit items.

10.1.3 Acceptance

Some of the general reactions that have been levied from the press and civil libertarians were described earlier. There are genuine concerns here, and it is important that we proceed cautiously to build in the necessary safeguards to protect us against RFID misuse. In 2003, one author proposed "An RFID Bill of Rights" [55] that laid down a set of guidelines that retailers should adhere to in order to protect the rights of our citizens. At present there are no laws regulating how tags can be used, and to gain full public acceptance legislation might be required. However, in the meantime, the early adopters such as Wal-Mart and Tesco (UK) could help defuse the current concerns by publicly adopting their own 'Bill of Rights' as an open policy.

10.2   ADDITIONAL CHALLENGES FOR SHORT-RANGE RFID

When describing short-range RFID applications, we often assume it is known where a tag has been placed on an object, so that a reader can be brought close-by for interrogation. However, an advantage of RFID is that it can be hidden in packaging and placed behind conventional labels, and therefore does not spoil the aesthetic appearance of the product. But invisibility also means that anybody unfamiliar with the product does not know where to look for an attached tag and thus where to place a reader. In the case of RFID tags that contain memory which may remain active throughout a product's lifetime, this is an important issue to resolve for the consumer as well as the manufacturer and retailer. However, a standardized solution to this problem has yet to be found [26]. One possible solution is to establish a convention for RFID placement. For example, it could always be located behind the manufacturer's logo, or in a location, such as the topside of the product. Alternatively there might be a discreet, but unmistakable, symbol that is placed in front of where the tag is embedded. In comparison to barcode technology, this symbol would be much smaller and attract less attention, but still allow a user to visually search for the symbol in order to position the reader.

10.3   FUTURE TECHNOLOGIES

Although conventional silicon chips bonded to spiral copper coils, or UHF dipoles etched into a copper/acetate substrate, can be cost reduced to the point where they are viable for the supply chain market, there will be commercial pressure to reduce costs further. However, ultimately the assembly, or encapsulation processes employed will constrain the minimum price of a tag. In order to overcome this limitation, Alien Technology Inc., has been experimenting with self-assembly techniques for joining the tag silicon with the antenna. Although still in an early stage of development, they have been able to use a fluidic assembly process to streamline

SEM photograph of
185 micron nanoblocks

185 and 70 micron
nanoblock circuits on top
of a dime

FIGURE 10.1: Self-assembly techniques being pioneered by Alien Technology Inc.

tag manufacture. To do this, the silicon substrate is formed into an inverted pyramidal shape, a nano-block (see Fig. 10.1), and then in quantity the nano-blocks are added to a solution allowing each one to move randomly until it finds a similarly shaped well in an intermediate substrate (*non*-silicon). The two pieces slot together, and the resulting combination is held in place by surface effects. From a manufacturing perspective the new structure is larger, easier to pick-up, and maneuver into place to bond with the antenna. As a result the cost of assembly will drop accordingly.

In an even bolder approach to lowering costs, Philips (Eindhoven, Netherlands) has been experimenting with an all plastic RFID tag [56], see Fig. 10.2a, the raw materials and manufacturing process being less costly than silicon. By depositing organic semiconductor materials directly onto an acetate substrate, it is possible to design and build arbitrary plastic circuits. Early active devices made from organic polymers were only able to switch at low frequencies (~100 kHz), but recent improvements have enabled a 13.56 MHz RFID tag to be built entirely out of plastic, and return a unique 64-bit code to the reader. The tag is made from a plastic called pentacene, and this material along with smaller dimensions used to build the active devices, has resulted in the necessary speed improvement. About 2000 transistors are used in the prototype (Fig. 10.2b). This is an important development because 13.56 MHz is the same frequency used

(a)

(b)

FIGURE 10.2: (a) Philips' experimental set-up for testing a plastic RFID Tag. (courtesy Philips Electronics, N.V.) (b) A prototype plastic RFID tag up-close (courtesy Philips Electronics, N.V.)

by the near-field based ISO standards 15 693 and 14 443, and creates an opportunity to use a large pre-installed reader base to exploit this new lower cost tag. This project has demonstrated significant progress, but there are still many problems to overcome before commercial tags can be mass-produced. For example, the prototype tag shown in Fig.10.2b can only transmit its ID to a reader over a few millimeters, and was built using a conventional lithographic process. The range needs to be increased to a few centimeters for effective use; and to be cost competitive, an ink-jet deposition process must be used to assemble the organic transistors. This will allow less expensive materials and a scalable manufacturing process to replace today's silicon based approach. The potential of the ink-jet process has been shown by other groups [57], and if perfected will enable the antenna to be deposited on a substrate at the same time as the electronics, thus not requiring any second stage assembly. The results presented by Philips at ISSCC'06 are nonetheless very encouraging and will spur other researchers to work in this area.

10.4   CONCLUSION

RFID is continuing to make inroads into inventory control systems and it is only a matter of time before the component costs fall below a point that, when weighed against the advantages, make it an attractive economic proposition. While engineering challenges still exist, there are extensive developments underway to build tag-reading systems that have sufficiently high accuracy to perform acceptably. There may even be some economic pressure from the larger distributors to modify product packaging to integrate RFID and improve the read accuracy. At this delicate stage, while the technology is being trialed by major corporations, media reaction, and outspoken privacy groups have the opportunity to influence the rules by which the technology is used. Given there is now legislation in place among most developed countries to protect personal information held in computers at banks and other commercial organizations, there is no reason why RFID data management cannot acquire a similar code of conduct [55]. The potential benefits of RFID are enormous, and as long as the use of tag data is handled appropriately, we are certain to see many novel and surprisingly useful applications of this technology in the future.

References

[1] Uniform Code Council, Inc., *UPC Symbol Specification Manual*, Dayton, OH, Reprinted May 1995.

[2] D. J. Collins and N. N. Whipple, *Using Bar Codes—Why It's Taking Over*. Data Capture Institute, ISBN 0-9627406-0-8.w.

[3] C. Wu, "Tagged out," Science News, in *Taggants: Barcodes for Bombs*, Sept. 14, 1996.

[4] K. Finkelzeller, *The RFID Handbook*, 2nd edn., Wiley, 2003, ISBN 0-470-84402-7.

[5] R. Want, "The magic of RFID," *ACM Queue Magazine*, vol. 2, no. 7, pp. 41–48, Oct. 2004.

[6] R. Want, "RFID: The key to automating everything," *Scientific American*, pp. 56–65, Jan. 2004.

[7] Alien Technology, Inc., www.alientechnology.com.

[8] Reed Electronics Group, "RFID tags and chips: Opportunities in the second generation," Rep. No. IN0502115WT, www.instat.com, Dec. 2005.

[9] R. Want and D. Russell, "Ubiquitous electronic tagging," *IEEE DS-Online*.

[10] D. J. Moore, R. Want, *et al.*, "Implementing phicons: Combining computer vision with infrared technology for interactive physical icons," in *Proc. ACM UIST'99*, Ashville, NC, pp. 67–68, Nov. 8–10, 1999.

[11] R. Want, A. Hopper, V. Falcao, and J. Gibbons, "The active badge location system," *ACM TOIS*, vol. 10, no. 1, pp. 91–102, Jan. 1992.

[12] J. Paradiso and M. Feldmeier, "A compact self-power push button controller," in *Ubicomp 2001*, Springer Verlag, Berlin, pp. 299–304, 2001.

[13] EPCglobal, Inc., www.epcglobalinc.org.

[14] IBM, AssetID, Information Brief, www.pc.ibm.com/ww/assetid/index.html, Nov. 1999.

[15] Indala, (Motorola), *BiStatix White Paper, Version 4.1*, www.mot.com/LMPS/Indala/bistatix.htm, March 1999.

[16] Microchip Technology, *RFID System Design Guides and Application Notes*, www.microchip.com, 1999.

[17] Trovan RFID, "Method and apparatus for modulating and detecting a sub-carrier signal for an inductively coupled transponder," www.trovan.com, US Patent #5095309.

[18] TIRIS, *Tag-it Inlays*, Product Bulletin, Texas Instruments, www.tiris.com.

[19]   W. H. Hayt, Jr., *Engineering Electromagnetics*, 5th edn., McGraw-Hill, New York, ISBN 0-07-027406-1.

[20]   G. Moore, "VLSI: Some fundamental challenges," *IEEE Spectrum*, vol. 16, p. 30, 1979.

[21]   Specification of Air Interface—EPCglobal. "EPC^{TM} Radio-Frequency Identity Protocols Class-1 Generation-2 UHF RFID Protocol for Communications at 860–960 MHz Version 1.0.9," Jan. 2005.

[22]   Impinj, Inc., www.impinj.com.

[23]   Near Field Communication (NFC) Forum, www.nfc-forum.org.

[24]   The Bluetooth Special Interest Group (SIG), www.bluetooth.org.

[25]   WiFi IEEE 802.11 wireless standard, http://standards.ieee.org/getieee802/802.11.html.

[26]   R. Want, K. Fishkin, B. Harrison, and A. Gujar, "Bridging real and virtual worlds with electronic tags," in *Proc. ACM SIGCHI*, Pittsburgh, pp. 370–377, May 1999.

[27]   M. Weiser, "The computer for the 21st century," *Scientific American*, vol. 265, no. 3, pp. 94–104, Sept. 1991.

[28]   A. Tannenbaum, *Computer Networks*, Prentice Hall, Englewood cliffs, NJ, 1981.

[29]   P. Karn, "MACA—A new channel access method for packet radio," in *Proc. 9th ARRL/CRRL Amateur Radio Computer Networking Conf.*, London, Ontario, Canada, p. 134, Sept. 22, 1990.

[30]   M. Alesky, *Cyborg 1.0*, *Wired Magazine*, Issue 8.02, pp. 144–151, Feb. 2000.

[31]   R. Want, "Enabling ubiquitous sensing with RFID," *IEEE Comput.*, vol. 37, no. 4, pp. 84–86, April 2004.

[32]   Microsensys, *Temperature Sensing and Logging*, www.microsensys.com/english/emicros.htm.

[33]   K. Opasjumruskit, *et al.*, "Self-powered wireless temperature sensors exploit RFID technology," *IEEE Pervasive Comput.*, pp. 54–61, Jan.–March 2006.

[34]   KSW Microtec, www.ksw-microtec.de.

[35]   D. E. Culler and H. Mulder, "Smart sensors to network the world," *Scientific American*, pp. 85–91, June 2004.

[36]   M. Philpose, J. R. Smith, B. Jiang, A. Mamishev, S. Roy, and K. Sundara-Rajan, "Battery-free wireless identification and sensing," *IEEE Pervasive Comput.*, vol. 4, no. 1, pp. 37–45, Jan.–March 2005.doi:10.1109/MPRV.2005.7

[37]   J. Smith, *et al.*, "A wirelessly powered platform for sensing and computation," In *8th Int. Conf. Ubiquitous Computing*, Orange Country, CA, USA, pp. 495–506, Sept. 17–21, 2006.

[38]   Metro Group Store Future Store Initiative, www.future-store.org.

[39]   Internet Retailer, www.InternetRetailer.com Article id=17448.

[40] C. Legner and F. Thiesse, "RFID-based maintenance at Frankfurt airport," *IEEE Pervasive Comput*, vol. 5, no. 1, pp. 34–39, Jan.–March 2006.doi:10.1109/MPRV.2006.14

[41] M. Philipose, K. Fishkin, D. Patterson, M. Perkowitz, D. Hahnel, D. Fox, and H. Kautz, "Inferring activities from interactions with objects," *IEEE Pervasive Comput.*, vol. 3, no. 4, 2004.

[42] J. R. Smith, K. P. Fishkin, B. Jiang, A. Mamishev, M. Philipose, A. D. Rea, S. Roy, and K. Sundar-Rajan, "RFID based techniques for human-activity detection," *Commun. ACM*, vol. 48, no. 9, pp. 39–44, Sept. 2005.doi:10.1145/1081992.1082018

[43] K. Fishkin, M. Philipose, and A. Rea, "Hands on RFID: Wireless wearables for detecting use of objects", in *Proc. 9th Int. Sym. of Wearable Computers, ISWC'05*, Oct. 2005.

[44] F. Thiesse, E. Fleisch, and M. Dierkes, "LotTrack: RFID-based process control in the semiconductor industry," *IEEE Pervasive Comput.*, vol. 5., no. 1., pp. 34–39, 47–53, Jan.–March 2006.doi:10.1109/MPRV.2006.14

[45] G. Borriello, W. Brunette, M. Hall, C. Hartung, and C. Tangney, "Reminding about tagged objects using passive RFIDs," in *6th Int. Conf. Ubiquitous Comput.*, Nottingham, UK, pp. 36–53, Sept. 2004.

[46] M. Ohkubo, K. Suzuki, and S. Kinoshita, "RFID privacy issues and technical challenges," *Commun. ACM*, vol. 48, no. 9, pp. 66–71, Sept. 2005.doi:10.1145/1081992.1082022

[47] E. Batista, "Step back' for wireless ID tech?" *Wired News*, http://www.wired.com/news/wireless/0,1382,58385,00.html, April 8, 2003.

[48] R. Singel, "American passports to get chipped," *Wired News*, www.wired.com/news/privacy/0,1848,65412,00.html, Oct. 19th, 2004.

[49] A. Juels, R. L. Rivest, and M. Szydlo, "The blocker tag: Selective blocking of RFID tags for consumer privacy," in *8th ACM Conf. Comput. Commun. Security*, ACM Press, pp. 103–111, 2003.

[50] P. Hewkin, "Smart tags—The distributed memory revolution," *IEEE Rev.* (UK), June 1989.

[51] T. Kindberg, *et al.*, "People, places and things: Web presence of the real world," *ACM MONET* (*Mobile Networks and Applications Journal*), 2002.

[52] P. Shih, V. Bahl, and M. J. Sinclair, "Wake on wireless: An event drive energy saving strategy for battery operated devices," in *Proc. ACM MobiCom 2002*.

[53] T. Pering, R. Ballagas, and R. Want, "Spontaneous marriages of mobile devices and interactive spaces," *Commun. ACM*, vol. 48, no. 9, pp. 53–59, Sept. 2005. doi:10.1145/1081992.1082020

[54] D. W. Engels, *The Reader Collision Problem*, Auto-ID Center White Paper, MIT-AUTOID-WH-007, Nov. 1, 2001.

[55]   S. Garfinkel, "An RFID Bill of Rights," *Technol. Rev.*, p. 35, Oct. 2002.

[56]   Philips demonstrates world-first technical feasibility of 13.56-MHz RFID tags based on plastic electronics. http://www.research.philips.com/newscenter/archive/2006/060206-rfid.html.

[57]   PARC Research, "Jet-printed plastic transistors—A solution for the display industry," http://www.parc.xerox.com/research/projects/lae/plastic.html.

Glossary

AFI	Application Family Identifier
ASK	Amplitude Shift Keying
BPSK	Binary Phase Shift Keying
CRC	Cyclic Redundancy Code
CSMA	Carrier Sense Multiple Access
DARPA	Defense Advance Research Program Agency
ECMA	European Computer Manufacturers Association
ECU	European Currency Unit
EPC	Electronic Product Code
ETSI	European Telecommunications Standards Institute
FCC	Federal Commission of Communication
HF	High Frequency
ID	Identification/Identity
ISO	International Standards Organization
ISM	Industrial Scientific Medical
JINI	A Discovery Service created by Sun Microsystems
LSB	Least Significant Bit
LF	Low Frequency
MACA	Media Access Collision Avoidance
MAC	Media Access Control
MSB	Most Significant Bit
NFC	Near Field Communication
PC	Personal Computer
PSK	Phase Shift Keying
RFID	Radio Frequency Identification
ROM	Read Only Memory
SNR	Signal Noise Ratio
UPC	Universal Product Code
UHF	Ultra High Frequency
UPnP	Universal Plug and Play

WiFi	IEEE 802.11a/b/g Standard
WLAN	Wireless Local Area Network
WORM	Write-Only, Read-multiple Memory
XOR	Exclusive-OR
16RN	16-bit Random Number

Author Biography

Roy Want is a Principal Engineer at Intel Research. Interests include embedded systems, mobile computing and automatic identification. Want received a BA in computer science from Cambridge University, UK in 1983 and earned a Ph.D. in distributed multimedia-systems in 1988.

He joined Xerox PARC's Ubiquitous Computing program in 1991 and managed the Embedded Systems area, later earning the title of Principal Scientist. He joined Intel Research in 2000.

Want is the author of more than 50 publications in the area of mobile and distributed systems; and holds 52 patents. He is a Fellow of both the IEEE and ACM.

Figure Acknowledgments

Figures 1.1, Altek Instruments Ltd, BarcodeMan, Walton on Thames. UK

Figure 1.5, data in table extracted from *RFID Tags and Chips: Opportunities in the 2nd Generation,* Report # IN0502115WT, Publisher, Reed Business

Figure 5.1, HID Global Corporation, Irvine, CA 92618 U.S.A www.hidcorp.com

Figure 5.2, Inspec Tech, Inc. Valley Head, Alabama www.inspectech.us

Figure 5.5b, The Augusta Chronicle, and the Boston Athletic Association

Figure 5.6a, North Dakota State University, Extension Service, Fargo, North Dakota

Figure 5.6b, Pubaa Animal Clinic, Segamat, Malaysia

Figure 5.9a, Metropolitan Transportation Commission Oakland, California www.mtc.ca.gov

Figure 6.1 and 6.2, KSW Microtec AG, Dresden, Germany, www.ksw-microtec.de

Figures 6.3, 10.2a, 10.2b, Philips Electronics, N.V., Eindhoven, The Netherlands

Figure 7.1, METRO Group Future Store Initiative, METRO AG, www.future-store.org

Figure 7.3, from IEEE, Pervasive Computing, Volume 5, No. 1

Figure 7.5, University of Washington, Seattle, Washington

Figure 8.2a, 8.2b, Notags.co.uk, UK "Citizens against the pervasive use of RFID in our Society"

Figure 9.2a, Centre for Orienteering History, www.orienteering-history.info

Figure 9.2b, Deeside Orienteering Club, UK, www.deeside-orienteering-club.org.uk

Figure 10.1a, 10.1b, 10.1c, Alien Technology Corporation, www.alientechnology.com

Printed in the United States
by Baker & Taylor Publisher Services